Physical Science:
What the Technology Professional Needs to Know
A Laboratory Manual

Dr. Richard Zajac

WILEY

JOHN WILEY & SONS, INC.

New York / Chichester / Weinheim / Brisbane / Singapore / Toronto

Technical Illustration: Richard J. Washichek, Graphic Dimensions, Inc.
Research: Judy Sullivan

This book is printed on acid-free paper. ∞

This material is based on work supported by the National Science Foundation under Grant No. DUE97-51988. Any opinions, findings, and conclusions or recommendations expressed in this material are those of the author(s) and do not necessarily reflect those of the National Science Foundation.

Library of Congress Cataloging-in-Publication Data:

ISBN: 978-0-4713-6019-3

Table of Contents

Preface

This laboratory manual is intended for use in conjunction with the text, *Physical Science: What the Technology Professional Needs to Know* or other appropriate book. It is designed for the student who has little or no background in the study of physics or chemistry, but is interested in improving their skills or preparing for a career as a technology professional.

If it is agreed that scientific progress is dependent on finding answers to problems through experimentation and interpretation of the data, then students should be given the opportunity to experience science using this method. Each experiment has therefore been designed to achieve this objective using the following format:

■ The focus of each experiment is developed using a background discussion. The discussion describes the overall experiment as well as reviewing important information and formulas.

■ Each experiment is subdivided into a series of smaller investigations. The objectives of each investigation are presented and a procedure outlined for collecting the necessary data. The data collected are then interpreted at the end of each investigation.

■ Preliminary interpretation of the data may involve calculations, graphical analysis, and/or spreadsheets. On occasion, graphing and spreadsheets are also used as predictive tools. The questions included in the interpretation help keep the student on target.

■ At the end of the last investigation, students are asked to move selected data and conclusions from each investigation to the Report Sheet. This allows to students to demonstrate their understanding and it narrows the information that the teacher must review and evaluate. Instructors may wish to have the students turn in the Report Sheets only, or staple them to the top of the other sheets in the experiment. The former allows the students to select their data and turn in only their best work; the latter would be useful for "a deeper analysis" if the Report Sheet signals a problem in one of the Investigations.

■ At the end of each Report Sheet, an extra credit section provides the student with an opportunity to further demonstrate their understanding by applying it to new situations.

■ Experiments are designed for completion within a three-hour period.

There has been an attempt to use only simple, inexpensive equipment to perform the experiments. The principles of "small scale" chemistry were adopted to keep the quantity of chemicals used to a minimum. Care has also been taken to avoid the use and later disposal of potentially hazardous chemicals by using "benign by design" chemicals whenever possible.

The Teacher's Guide accompanying this laboratory manual is provided to assist in the planning and evaluation of the laboratory experience. An overall discussion of each experiment's objectives is followed by a brief discussion of student expectations and pitfalls while performing the various investigations. A list of expected foundational skills to facilitate student performance is given along with lists of the equipment and supplies required. Directions are provided for the preparation of some reagents as well as sources for specialized equipment. A highlighted box listing the safety precautions specifically related to the performance of the experiment are included. Finally, typical student data and conclusions are given for each of the investigations and the Report Sheet to assist the teacher in evaluating student performance.

Acknowledgments

The author wishes to acknowledge the following organizations and individuals that have made this laboratory manual possible:

- INTELECOM, Intelligent Telecommunications and the National Science Foundation (NSF) for making and funding this project possible.

- The Partnership for Environmental Technology Education (PETE) who nominated the regional representatives to serve on the National Academic Council (NAC).

- The NAC members for their help in developing and designing the project:
  - Ann Boyce, Bakersfield College;
  - Eldon Enger, Delta College;
  - Doug Feil, Kirkwood Community College;
  - Christine Flowers, Shasta College;
  - Ross Marano, El Centro Community College;
  - A. J. Silva, South Dakota School of Mines;
  - Larry Stewart, Highland High School; and
  - Brent Wurfel, Arkansas State University – Mountain Home.

- Graphic Dimensions for their talent in converting the manuscript and graphics into an attractive and useful manual.

- Judy Sullivan, INTELECOM Editor, for her numerous contributions, skills, and handling of the many project details.

- Howard Guyer, Academic Team Leader, for his contributions, review of the experiments, and for his tenacity and vision throughout the entire process.

Dr. Richard Zajac
Kansas State University – Salina

Note to the Student

The laboratory is the heart of science. Textbooks can use only words to describe color changes, smells, and the feel of the equipment in a science laboratory. The selection of experiments for this manual has been built around major physics and chemistry concepts. Each experiment is then broken into a few simpler investigations, each with its own set of objectives. The interpretation of data section at the end of each investigation will help you focus on the conclusions aid in understanding the objectives. The various sections in each experiment and their purpose are the following:

■ Background Concepts: The background is intended to focus the area of study and review those concepts and formulas that are necessary to perform and understand the experiment.

■ Investigations: Each experiment is subdivided into three or four investigations. Each investigation has its own set of objectives, procedures, and interpretation. It is the purpose of the objectives to narrow the focus. The procedure directs the activities you should follow and data that is to be collected. The interpretation, through a series of questions and activities, directs the calculations, graphing, or spreadsheets to be develop to answer the questions and reach the objective.

■ Report Sheet: The Report Sheet gives you the opportunity to present your "best" work. Errors that may have initially been made and later corrected need not be shown on this sheet. The "extra credit" portion of each Report Sheet provides you with an opportunity to demonstrate your mastery of the experiment.

The experiments provided in this manual are ones that can be performed by any student who is willing to follow a few directions. Your understanding and performance time will be greatly enhanced if you a few minutes to read the entire experiment prior to the laboratory period. If you are not thoroughly familiar with a concept discussed in the experiment, review it in your textbook prior to the lab.

Follow all safety precautions, both those given by your teacher and in print. Carefully dispose of chemicals as directed by your instructor. Do not use your laboratory time to socialize. Keep the atmosphere in the lab business like and friendly. Move carefully through the isles taking care not to bump a fellow student or their experiment. It is generally permissible to communicate quietly with your fellow students, but it is always unacceptable to yell at someone across the room.

Welcome to the heart of science – the laboratory, where discoveries are made.

Experiment/Text Correlation and Student Objectives

Exp. No.	Exp. Title	Text Ch.	Student Objectives
1	Concrete: A Common Mixture	2	Read volumes from a graduated cylinder and use a centigram balance. Determine void volumes and bulk densities. Work with mixed units and calculate percents.
2	Density/Buoyancy Relationships	2	Make accurate measurements using a balance, metric rule, and graduated cylinder. Prepare and use graphs to average data and make predictions. Explain the concepts of density, specific gravity, and buoyancy.
3	Uncertainty, Error Bars, and Calibration	3	Explain how uncertainty arises in measurement. Demonstrate how a measurement's uncertainty range can affect further calculations based on the measurement. Explain the relationship between uncertainty and reproducibility.
4	Percent Composition and Error Analysis	4	Determine the percent of water in hydrated substances. Calculate the mean value of a series of measurements. Calculate the percent of error in an experimentally determined value.
5	Estimating the Atomic Mass of Metals (Law of Dulong & Petit)	4	Perform basic calorimetry studies. Determine the mathematical relationship between the specific heat and atomic mass of metals. Use the relationship discovered to determine the approximate atomic mass of an unknown metal.
6	Using Spreadsheets to Analyze Objects in Motion	5	Distinguish motion with a constant velocity from motion with a constant acceleration. Compute distances and time for simple motions. Present motion using graphical representations. Prepare spreadsheets to graphically represent and analyze data.
7	Objects in Motion	5	Distinguish motion with a constant velocity from motion with a constant acceleration. Compute distances and time for simple motions. Present motion using graphical representations.
8	Momentum and Friction in a Car Crash: A Forensic Investigation	5	Calculate the momentum of a moving object. Explain how the total momentum of two objects is the same before and after they collide. Calculate the coefficient of static friction between two objects.
9	Waves and Oscillations	5	Determine a wave's period of oscillation. Explain how the frequency and period of an oscillation are related. Demonstrate how a wavelength is measured. Visualize a traveling longitudinal wave and explain its connection to an oscillation. Determine the wave velocity. Explain how the size and end conditions of a resonant cavity affects the standing waves it can support.
10	Simple Machines	5	Explain how forces can be modified by simple machines; Calculate a mechanical advantage based on practical measurements; Explain what is meant by and calculate the efficiency of a simple machine.
11	Gas Laws	6	Determine the relationship between the change in the length of an air column and absolute temperature, measured at constant pressure. Determine the relationship between the volume of air trapped and the length of an air column. Determine the relationship between the volume and absolute temperature of an air sample, at constant pressure
12	Energy	6	Recognize the different forms of energy, including gravitational, kinetic, electrical, thermal, and chemical. Calculate the conversion of energy from one form to another.

Exp. No.	Exp. Title	Text Ch.	Student Objectives
13	Heat of Reaction (Hess' Law)	7	Determine if the lattice energy or heat of solvation is greater when an ionic substance dissolves in water. Measure the heat of reaction, Q, and calculate the enthalpy, ΔH, for several different kinds of reactions. Confirm Hess' Law by experimentally measuring a predicted enthalpy.
14	Exploration of Acids and Bases	7	Determine the effect of acidic and basic solutions on indicators. Determine if a solution is acidic or basic, based on its chemical characteristics. Calculate the amount of energy produced by an acid/base neutralization reaction.
15	Acid Concentrations and Strengths	7	Explain the difference between concentration and weak and strong acids. Discuss the difference between a monoprotic and diprotic acid. Describe the relationship between acid strength, concentration, and pH. Calculate the volume of base required to neutralize a known volume and concentration of acid.
16	Percent of Acetic Acid in Vinegar: An Acid/Base Titration	7	Calibrate the volume of a drop. Prepare a standard solution. Standardize an unknown solution using a standard solution. Use a standardized solution to determine some property of another substance.
17	Build Your Own Voltmeter	8	Explain how the voltage changes along a series circuit. Discuss why the current remains constant throughout a series circuit. Design a simple analog voltmeter.
18	Build Your Own Ammeter	8	Describe how current is divided in a parallel circuit. Explain how parallel branches experience the same voltage drop. Design a simple analog ammeter.
19	Refraction	8	Explain how the index of refraction is used to describe and calculate angles of refraction of a light ray in different materials. Explain how the change in direction of a light beam arises from a change in the speed of light from one material to another.
20	Diffraction Gratings	9	Explain how light waves can interfere constructively and destructively. Explain how diffraction gratings separate light into its component colors. Determine the wavelength of a laser beam.
21	Optics of Thin Lenses	9	Explain the difference between real and virtual images. Calculate focal lengths of lenses based on image and object distances. Calculate the final image location when multiple lenses are used. Explain how a lens can have a virtual object.
22	Spectrophotometry	9	Operate a spectrophotometer. Identify the major absorbance peak(s) of a colored solution. Identify lambda $_{max}$, λ_{max}, of a colored solution. Predict the color of a solution from the absorbance peak(s) and λ_{max} data.
23	Molecular Models	10	Explain the difference between a molecular, full structural, and condensed structural formula; Write the general formulas for the alkane, alkene, and alkyne families; Demonstrate a method for determining the number of isomers for a simple molecular formula.
24	Organic Esters	10	Make several artificial flavorings. Prepare aspirin. Describe some of the properties of PET plastic.
25	Using Properties to Identify Organic Families	10	Determine the water solubility of organic compounds. Determine if an organic compound is saturated or unsaturated. Discuss why some water soluble organic families result in acidic and others basic solutions.
26	Simulating Nuclear Processes	11	Explain how statistical processes can produce exponential behavior. Calculate decay constants and half-life from a given decay.

Is there a difference between cement and concrete?

Concrete: A Common Mixture

Background Concepts

Sometimes things turn out to be different than we realize. For instance, we tend to use the terms cement and concrete interchangeably, but a closer examination reveals that they are not the same. Portland cement is an ultra-fine gray powder that binds sand and rocks into a mixture called concrete. It was named after a similar looking stone quarried on the Isle of Portland. The iron and manganese compounds present are what give it a gray color.

Concrete is typically composed of 60 to 75% aggregate, by volume, that is a mixture of sand, gravel, and/or crushed rock. In recent years, the industry has been increasing the percent of recycled concrete used as aggregate. The only real requirements for the aggregate are that it be clean, hard, strong, and free of chemicals that would interfere with the chemical hydration processes occurring during the curing process.

Concrete is used for highways, pipes, buildings, bridges, and a host of other applications. It is the most widely used building material in the world; in fact, there are those who say that we are moving into a new Stone Age! Current annual global production of concrete is about 5 billion cubic yards, which requires the production of over 1.25 billion tons of cement.

Since the first civilizations started to build, there has been a need for a substance that would bind stones into a solid mass. The Assyrians and Babylonians used clay. The Egyptians used a lime and gypsum mortar to build such structures as the Pyramids. The Romans developed a type of cement that was used as the foundation for the Roman Forum. The great Roman baths built about 27 BC, the Coliseum, and the huge Basilica of Constantine are other examples of early Roman architecture in which cement mortar was used. The secret of Roman concrete making was eventually traced to the mixing of slaked lime with volcanic ash from Mount Vesuvius. This combination produced hydraulic concrete – concrete that would harden under water. During the Middle Ages the art was lost, but was later rediscovered.

Today, we understand that concrete is little more than a mixture of rocks and paste. The rocks are typically a combination of sand (small rocks) and larger rocks and the paste is a combination of Portland cement and water. During mixing, the aggregate becomes coated with paste. A hydration process called curing follows, during which the mixture hardens into a rocklike mass called concrete.

The most important factor in determining the strength of concrete is the ratio of water to cement. The higher the ratio of cement, the stronger the concrete becomes in every desirable physical property. The easiest way to strengthen concrete, therefore, is to add more cement, but that makes it more expensive. A less expensive way to produce strong concrete is to reduce the amount of void space between the aggregate particles so less cement is needed.

In this lab, you will explore the amount of void space between different sized aggregate particles and mixtures of aggregate particles. From this, you will be able to determine an ideal ratio of aggregate and cement to produce the strongest concrete using the least cement.

Investigation 1

Objective

To determine the void percent and bulk density of a small aggregate, sand.

> ## Safety Requirements
>
> ■ Always wear splash proof safety goggles.
>
> ■ Discard the waste materials as directed by your instructor.
>
> ■ Portland cement is caustic to the skin. Wash exposed skin thoroughly with cold water and soap.
>
> ■ Return "used" aggregate to collection containers as directed by your instructor.

Procedure

1. Obtain a dry plastic cup and determine its mass to the nearest ± 0.1 g. Record the mass of the cup in Table 1-1.

2. Make a mark near the top of the cup. Fill the cup to the mark with tap water.

Cylinder Reading #1	mL
Cylinder Reading #2	mL
Cylinder Reading #3	mL
Total Volume of Cup to Mark	mL
Mass of cup and sand	g
Mass of cup	g
Mass of Sand	g
Volume of H_2O added	mL
Volume of H_2O added	mL
Total Volume of H_2O Used	mL
Percent Void Volume of Sand	%
Bulk Density of Sand	g/mL

Table 1–1: Determination of the void percent and bulk density of sand.

3. Measure the volume of water in the cup by first pouring it into a 400 mL beaker and then transferring it from the beaker into a 100 mL graduated cylinder. Each time the water level in the cylinder nears the 100 mL mark, carefully read and record the volume in Table 1-1. Empty the cylinder and continue this process until the cup is empty.

4. Dry the cup and fill it to the mark with dry sand. Eliminate any large spaces between the small aggregate particles by gently bouncing it on the desk several times. Add additional sand, if necessary, to again fill the cup to the mark.

5. Determine the mass of the cup and sand to the nearest ± 0.1 g. Record the mass in Table 1-1.

6. Fill the graduated cylinder to near the 100 mL mark with cold tap water, read and record the volume in Table 1-1.

7. Slowly pour the water from the cylinder into the cup of sand until the water level is again at the mark near the top of the cup. Repeat steps 6 and 7, if necessary.

8. Determine and record in Table 1-1 the total volume of water required to bring the water level to the mark.

9. Using the method suggested by your instructor, separate the sand-water mixture or return it to the "used sand" container.

Interpretation

1. Calculate the volume of the cup to the mark and record it in Table 1-1.

2. Calculate the total mass of sand in the cup and record it in Table 1-1.

3. Calculate the volume of water used to fill the cup of sand to the mark and record it in Table 1-1. This is the void volume of the aggregate or the space remaining between the grains of sand per cup.

4. Calculate the percent void volume by dividing the volume of water required to fill the cup by the total volume of the cup times 100. Show your work and record the percent void volume in Table 1-1.

5. The bulk density of a material is defined as its mass divided by its bulk volume, including void space, $\rho = m/v$. Calculate the bulk density of the sand. Show your work and record your results in Table 1-1.

Investigation 2

Objective

To determine the void percent and bulk density for a large aggregate; rock.

Procedure

1. Use the same dry plastic cup that was used in Investigation 1. Fill it to the mark with large aggregate.

2. Use the same method as was used in steps 4 – 8 in Investigation 1 to determine the mass of large aggregate and amount of water required to fill the cup to the mark. Record the information in Table 1-2 and transfer any data necessary to complete the table from Table 1-1.

3. Using the method recommended by your instructor, separate the aggregate-water mixture or return it to the "used large aggregate" container.

Total Volume of Cup to Mark	mL
Mass of cup and large aggregate	g
Mass of cup	g
Mass of Large Aggregate	g
Volume of H_2O	mL
Volume of H_2O	mL
Total Volume of H_2O Used	mL
Percent Void Volume of Large Aggregate	%
Bulk Density of Large Aggregate	g/mL

Table 1–2: Determination of the void percent and bulk density of large aggregate.

Interpretation

1. Calculate the percent void volume by dividing the volume of water required to fill the cup of large aggregate to the mark by the total volume of the cup times 100. Show your work and record the percent in Table 1-2.

2. Calculate the bulk density ($\rho = m/v$) of the large aggregate. Show your work and record your results in Table 1-2.

3. Based on your findings in Investigations 1 and 2, what conclusions can you draw about the relationship between the percent of void volume and aggregate particle size?

Investigation 3

Objectives

To determine the void percent and bulk density of an aggregate mixture.

Procedure

1. Use the same dry plastic cup and again fill it to the mark with large aggregate.

2. Obtain more of the dry sand used in Investigation 1 and carefully start adding it to the cup full of large aggregate. You may need to again bounce the cup on the desk several times to help the grains of sand move downward into the cup. Continue this process until no more sand can easily be added.

3. Determine the mass of the cup and combined aggregate to the nearest ± 0.1 g. Record the mass in Table 1-3.

4. Fill the graduated cylinder to near the 100 mL mark with cold tap water, read and record the volume in Table 1-3.

5. Slowly pour the water from the cylinder into the cup of combined aggregate until the water level is again at the mark in the cup.

6. Determine the total volume of water required to fill the cup to the mark. Read and record in Table 1-3.

7. Using the method recommended by your instructor, separate the sand-aggregate-water mixture or return it to the designated "used mixed aggregate" container.

Total Volume of Cup to Mark	mL
Mass of cup and mixed aggregate	g
Mass of cup	g
Mass of Mixed Aggregate	g
Volume of H_2O	mL
Volume of H_2O	mL
Total Volume of H_2O Used	mL
Percent Void Volume of Mixed Aggregate	%
Bulk Density of Mixed Aggregate	g/mL

Table 1–3: Determination of the void percent and bulk density of mixed aggregate.

Interpretation

1. Transfer from Table 1-2 to Table 1-3 the total volume of the cup to the mark.

2. Calculate the percent void volume by dividing the volume of water required to fill the cup of aggregate mixture to the mark by the total volume of the cup times 100. Show your work and record the percent in Table 1-3.

3. Calculate the bulk density ($\rho = m/v$) of the aggregate mixture. Show your work and record your results in Table 1-3.

4. Based on your findings in the three investigations, which aggregate or combination of aggregates resulted in the lowest void volume percent? Explain.

5. Based on your findings in the three investigations, what aggregate or combination of aggregates would result in the strongest, least expensive, concrete. Explain your reasoning.

Investigation 4

Objective

To make a cup of concrete and determine its bulk density.

Procedure

> **Note**: To make strong concrete, experience has lead to the rule of 6's. This rule, in part, recommends the use of 6 gallons of water per bag of cement and a minimum of 6 bags (94 lb/bag) of cement per cubic yard of concrete. Based on the above rule, two ratios have been calculated and added to Table 1-4.

1. What is the volume of the cup, to the mark, that you have used throughout the previous investigations? Record that volume in Table 1-4.

2. What is the bulk density of the aggregate or aggregate mixture that had the least void volume? Record its bulk density in Table 1-4.

3. Calculate the number of grams of the aggregate or aggregate mixture that would be required to fill the cup to the mark. Show your work and record the amount in Table 1-4.

4. Using the mass ratio information in Table 1-4, calculate the number of grams of cement that is required by the Rule of 6's. Show your work and record the amount in Table 1-4.

Mass Ratio of Aggregate to Cement	5.41 to 1.00
Mass Ratio of Water to Cement	0.0638 to 1.00
Volume of Cup to Mark	mL
Bulk Density of "Best" Aggregate	g/mL
Mass of "Best" Aggregate to Fill Cup	g
Mass of Cement Required per Ratio	g
Mass of Water Required per Ratio	g
Mass of Wet Concrete	g
Mass of Cured Concrete (optional)	g
Mass Difference Wet and Cured Concrete (optional)	g
Filling Observations:	

Table 1–4: Typical concrete ratios based on the rule of 6's.

5. Using the mass ratio information in Table 1-4, calculate the number of grams of water that is required by rule of 6's. Show your work and record the amount in Table 1-4.

6. Using the approximation that $1.00 \text{ g} = 1.00 \text{ mL}$ of water, weigh out the calculated amounts of aggregate and cement in separate containers. Use your graduated cylinder to measure the required amount of water. Using your 400 mL beaker and a spatula as the "cement mixer," add the measured amounts of aggregate, cement, and water. Stir thoroughly to produce concrete with a uniform texture.

7. Carefully transfer the wet concrete into the cup. Eliminate spaces in the cup by gently bouncing it on the desk several times. Did the mixture fill the cup to the mark or not? Record your observation in Table 1-4.

8. Determine the mass of the cup and wet concrete. Record this value in Table 1-4.

9. (Optional) As directed by your instructor, set the cup of wet concrete aside until the next laboratory period, when you will again determine its mass and make observations.

10. Wash your beaker and spatula as directed by your instructor. Do not put any left over concrete in the sink or down the drain. Wash your hands prior to leaving the laboratory.

Interpretation

1. How nearly did the calculated and mixed concrete meet the target volume of the mark on the cup? Explain any differences.

2. What is the difference between cement and concrete? Explain your answer.

3. (Optional) Is there any difference between the masses of the "wet" concrete and the cured concrete one week later? Explain any similarities or differences.

4. (Optional) Remove the cured concrete from the cup. In mixtures, pure substances retain their original properties. After closely examining the cured concrete, are the original properties of any of its components still visible? Explain your observations.

Report Sheet

Experiment 1

1. Complete the following table with the information from Table 1-1.

Total Volume of Cup to Mark		mL
Mass of Sand		g
Total Volume of H_2O Used		mL
Percent Void Volume of Sand		%
Bulk Density of Sand		g/mL

2. Complete the following table with information from Table 1-2.

Total Volume of Cup to Mark		mL
Mass of Large Aggregate		g
Total Volume of H_2O Used		mL
Percent Void Volume of Large Aggregate		%
Bulk Density of Large Aggregate		g/mL

3. Complete the following table with information from Table 1-3.

Total Volume of Cup to Mark		mL
Mass of Mixed Aggregate		g
Total Volume of H_2O Used		mL
Percent Void Volume of Mixed Aggregate		%
Bulk Density of Mixed Aggregate		g/mL

4. Which of the aggregates or combination of aggregates resulted in the lowest void volume? Explain.

5. Complete the following table with information from Table 1-4.

Volume of Cup to Mark		mL
Bulk Density of "Best" Aggregate		g/mL
Mass of "Best" Aggregate to Fill Cup		g
Mass of Cement Required per Ratio		g
Mass of Water Required per Ratio		g
Mass of Wet Concrete		g
Mass of Cured Concrete (optional)		g
Mass Difference Wet and Cured Concrete (optional)		g

6. It is common to hear the terms "cement sidewalk," "a cement mixer" and "a cement truck." Based on this experiment, write a short explanation for why these terms are incorrect.

Extra Credit

1. The composition of concrete is often varied depending on its application. The following table contains average values for concrete used for purposes such as sidewalks and home foundations. Using the necessary information given, calculate how many pounds of concrete are in a typical 8 yd^3 truck load. Show your work.

Average Values	
Item	**Value**
Density of water	1.00 g/mL
Density of Portland Cement	3.51 g/cc
Density of Aggregate	2.75 g/cc
Concrete	140 lb/ft^3
Cement	12.5% of concrete, by volume
Aggregate	75% of concrete, by volume

What floats your boat?

Density-Buoyancy Relationships

Background Concepts

Have you ever had a problem, studied it, and attempted to find an explanation that would predict its solution? This is what science is all about, but scientific predictions have a feature that distinguishes them from all others: when properly used, they must enable anyone to make the same accurate prediction.

Since the beginning of time, humans have attempted to understand the properties of matter. Some of these properties had practical implications that were also of interest to craftsmen. For example, ship builders needed to know how much cargo could be loaded on a ship without danger of it sinking when it sailed from sea water into fresh water. The problem was solved by the Greek mathematician and physicist Archimedes who studied and explained the relationship between a fluid's density and its buoyant force. Archimedes' principle can predict not only the load a ship can safely carry, but the size a hot air balloon must be in order to lift a man from the ground.

Today we continue to find applications for Archimedes' principle. In many industrial and medical settings there is a need to monitor, either periodically or constantly, the density of solutions.

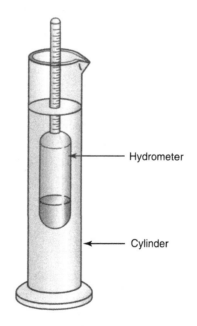

Figure 2–1: A hydrometer floating in a liquid.

The labels in the figure read: Hydrometer, Cylinder.

Time-consuming laboratory methods may be used, but thanks to Archimedes' principle, a more efficient device called a hydrometer is typically used. Hydrometers tend to float in a liquid at a depth that is directly proportional to the density of the liquid. Some everyday examples of where hydrometers are used include determining the amount of butterfat in milk (lactometers), alcohol in a beverage (alcoholometer), brine in a ship's boilers (salinometer), acid in a car's battery (acidimeter), and the specific gravity of urine (urinometer).

Hydrometers are typically made of glass and have some type of ballast material in the bottom. As shown in Figure 2-1, when placed in a liquid, the weighted lower end sinks leaving a portion of the stem above the liquid's surface. By using a series of liquids with known densities, the hydrometer's stem can be calibrated to give density readings directly. For example, if the hydrometer sinks 6.00 cm in water but 8.00 cm in alcohol, then the alcohol must have a density (6.00 cm/8.00 cm) 0.75 times that of water.

If the density of a liquid and the density of water are compared, the units divide out and the resulting number is called a specific gravity (sp gr). It is common practice to use specific gravity because it eliminates the need to include the units. As shown below, when the density of an alcohol solution is compared to the density of water, the units g/mL divide out.

$$\text{specific gravity} = \frac{\text{density of alcohol}}{\text{density of water}}$$

$$\text{specific gravity} = \frac{0.75 \text{ g/mL}}{1.0 \text{ g/mL}}$$

$$\text{sp gr} = 0.75$$

As in many scientific investigations, the larger problem – why hydrometers work – can often be subdivided into a series of smaller problems. Investigation 1 will introduce you to the basic concepts of density, specific gravity, and buoyancy. Once you have explored these relationships, you will be able to explain the technology of using a hydrometer as well as what floats your boat!

Investigation 1

Objective

To state a hypothesis that explains the difference between the number of spheres required to sink the tube in water and in salt water.

Procedure

1. Firmly place a cork into one end of a soda straw. See Figure 2-2.

2. Put 90 mL of deionized (DI) water into a 100 mL graduated cylinder.

3. Check the seal of the cork by pushing the straw down into the water to a depth of 15 cm. If the cork leaks, press the cork in tighter and repeat the test.

4. Float the tube in the graduated cylinder.

5. Count and record the number of spheres that must be added to make the tube touch the bottom of the cylinder.

   _____ spheres

6. Empty the graduated cylinder and fill it with approximately 90 mL of the salt solution provided.

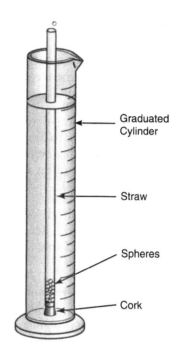

Figure 2–2: Cork in end of soda straw.

7. Float the same tube in the graduated cylinder of salt solution.

8. Record the number of spheres that must be added to make the tube touch the bottom of the cylinder.

   _____ spheres

Interpretation

1. Was the number of spheres required to sink the tube the same for both liquids? Are the results what you expected? Explain your answer.

2. How do the number of spheres recorded in steps 5 and 8 compare?

3. Suggest a possible explanation (hypothesis) for the number of spheres required to sink the tube.

Investigation 2

Objective

To develop a method for predicting the length that a tube will sink below the surface of a liquid when the total mass being buoyed is known.

Procedure

1. Using the balance, determine the mass of the tube and record it to the nearest ± 0.01 g in Table 2-1.

2. Using the balance, determine the mass of 20 spheres to be used as ballast. Calculate the average mass/ sphere to the nearest ± 0.01 g and record in Table 2-1.

3. Fill the graduated cylinder with 85-90 mL of DI water. Record the actual volume in Table 2-1 to the nearest ± 0.2 mL.

4. Put 2 of the ballast spheres in the tube and float it in the graduated cylinder. While keeping the tube away from the cylinder wall, read and record to the nearest ± 0.2 mL the water level in the graduated cylinder.

5. Using the metric rule, measure to the nearest millimeter and record the depth that the tube sinks below the surface of the water. Try to keep the tube away from the cylinder wall when making all measurements.

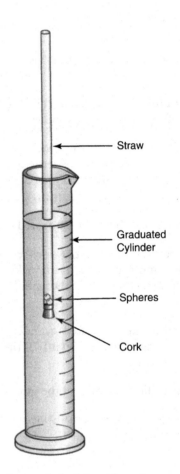

Figure 2–3: Spheres in soda straw.

6. Using the previous measurements, calculate the total mass being buoyed and add it to Table 2-1.

7. Read and record the level of the water after floating the tube and calculate the volume of water displaced. The volume of water displaced is the difference between the initial and present water level readings.

8. Add 1 more sphere to the tube. Record the water level in the graduated cylinder. Again, calculate the total mass being buoyed and the volume of water being displaced.

9. Repeat step 7 two more times and add this data to Table 2-1.

Mass of tube _____

Mass of 20 spheres _____

Calculated average mass/sphere _____

Initial volume of water
in graduated cylinder _____

Number of spheres added	Total mass buoyed (spheres + tube), g	Predicted length of tube below surface, cm	Actual length of tube below surface, cm	Volume of water after adding tube, mL	Volume of water displaced, (V with tube – V initial) mL
2		– –			
3		– –			
4		– –			
5		– –			
7					
9					
11					

Table 2–1: Investigation 2 data.

10. Graphs are convenient tools for finding regularities in data and making predictions. Select a convenient scale and construct Graph 2-1 using the data collected in Table 2-1. Use the horizontal or *x*-axis to show total mass buoyed and the vertical or *y*-axis for the actual length of the tube below the surface of the water. Identify each point through the use of a Δ or O. Draw the best representative line through the points. If it is a straight line, use a ruler.

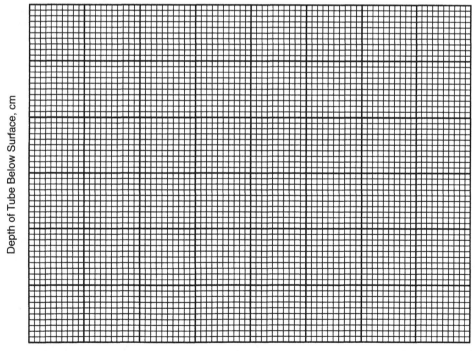

Total Mass Buoyed, g

Graph 2–1: Plot of depth and total mass buoyed.

11. By using the graph you prepared in step 10, predict the length of the tube that will sink below the surface of the water when 7, 9, and 11 spheres are used. Each time, record your prediction first and then check that prediction by adding that number of spheres and measuring the depth. Record all measurements and calculations in Table 2-1.

Interpretation

1. As the total mass being buoyed is increased, what effect does this have on the length of the tube being submerged?

2. As the total mass being buoyed is increased, what effect does this have on the volume of water being displaced?

3. What is the shape of the curve in Graph 2-1?

4. What relationship must exist between the total mass buoyed and the depth of submersion of the tube in the water to produce this shape of curve?

Investigation 3

Objective

To develop a method for predicting the length of the tube that will sink below the surface of a salt solution when the total mass is known.

Procedure

1. Using a salt solution, repeat the steps used in solving the previous problem. Record your data and calculations in Table 2-2.

2. Plot the data from Table 2-2 on Graph 2-1.

3. Use Graph 2-1 to make predictions concerning the length of the tube that will sink below the surface of the salt solution. Test your predictions and record them in Table 2-2. If your predictions of the length of the tube that will sink below the surface are not close, see your instructor.

Mass of tube _____ Mass of 20 spheres _____ Calculated average mass/sphere _____			Initial volume of salt solution in graduated cylinder _____		
Number of spheres added	Total mass buoyed (spheres + tube), g	Predicted length of tube below surface, cm	Actual length of tube below surface, cm	Volume of salt solution after adding tube, mL	Volume of salt solution displaced, (V with tube – V initial) mL
2		– –			
3		– –			
4		– –			
5		– –			
7					
9					
11					
15					

Table 2–2: Investigation 3 data.

Interpretation

1. Using the interpretation questions at the end of Investigation 1 as a guide, explain any similarities and differences in the results obtained when using water and salt solution as the buoying medium.

■ Investigation 4

Objective

To determine the relationship that exists between the total mass being buoyed and the volume of the liquids displaced.

Procedure

1. Fill in Table 2-3 extracting the necessary data from Tables 2-1 and 2-2.

Number of spheres	Total mass buoyed (spheres + tube), g	Volume of H_2O displaced, mL	Volume of salt solution displaced, mL	Length of tube below surface (H_2O), cm	Length of tube below surface (salt solution), cm
2					
3					
4					
5					
7					
9					
11					
15					

Table 2–3: Investigation 4 data.

2. Construct Graph 2-2 from the data in Table 2-3 for each of the solutions. The length of the tube submerged should be placed on the x-axis and the y-axis should be used to show the volume of the liquid displaced.

3. Construct Graph 2-3 from the data in Table 2-3 for each of the liquids. Use the x-axis to show the total mass buoyed and the y-axis to show the volume of the liquid displaced.

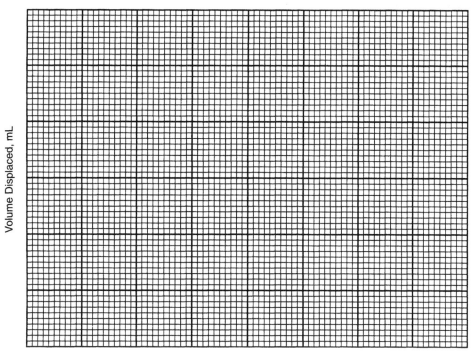

Length of Tube Submerged, cm

Graph 2–2: Volume of liquids displaced and length of tube submerged.

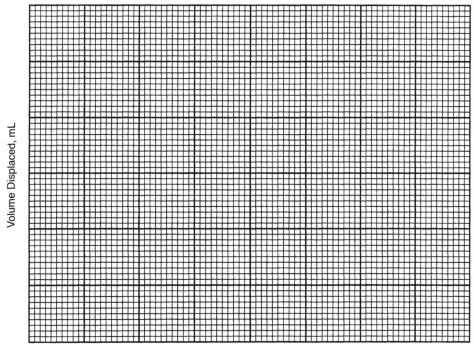

Total Mass, g

Graph 2–3: Volume displaced and total mass buoyed.

Interpretation

1. On Graph 2-3, find the mass being buoyed by the displacement of 1.0 mL of water.

2. Using Graph 2-2, determine the length of the tube submerged in water when 1.0 g is buoyed.

3. On Graph 2-3, find the mass being buoyed by the displacement of 1.0 mL of salt solution.

4. Using Graph 2-2, determine the length of the tube submerged in the salt solution when 1.0 g is buoyed.

5. What generalizations can be drawn based on the above findings?

Investigation 5

Objective

To find a property of liquids that may account for the differences in masses buoyed by the displacement of equal volumes of different liquids.

Procedure

The spheres, the liquids, the tube, and the graduated cylinder used in these investigations are all matter. They all occupy space and have mass. Attempt to determine the relationship between the mass and volume of the liquids used, by performing the following steps.

1. Determine and record the mass of a 10 mL graduated cylinder to the nearest ± 0.01 grams.

2. Using a dropper, place between 4 and 5 mL of water into the 10 mL graduated cylinder. Read and record in Table 2-4 the actual volume of water used to the nearest 0.1 mL.

3. Determine the mass of the cylinder and water. Record the data in Table 2-4.

4. Repeat step 2 using between 9 and 10 mL of water in the 10 mL graduated cylinder. Read and record in Table 2-4 the actual volume of water used to the nearest 0.1 mL.

5. Repeat steps 2, 3, and 4 using the salt water solution, and record this data in Table 2-4.

Liquid used	Actual volume of liquid used, mL	Mass of graduated cylinder, g	Mass of graduated cylinder + liquid, g	Mass of liquid, g
Water, DI				
Water, DI				
Salt Water				
Salt Water				

Table 2–4: Investigation 5 data.

6. Construct Graph 2-4 using the actual volumes of water and mass of water data collected in Table 2-4. Using a ruler, draw the best straight line through the points and origin.

7. Add the data collected in Table 2-4 for the salt solution to Graph 2-4.

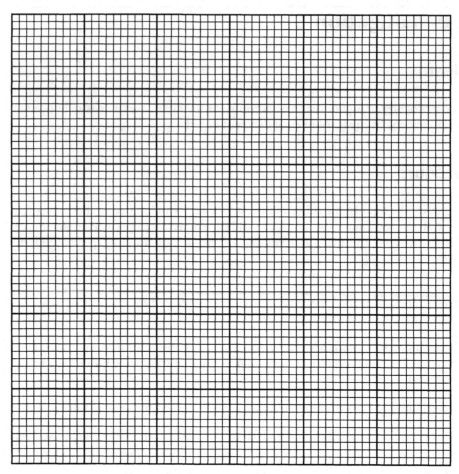

Mass, g

Graph 2–4: Densities of water and salt solution.

Interpretation

1. Determine from Graph 2-4 the mass of 1.0 mL of water.

2. Compare the mass of 1.0 mL of water to the mass buoyed by the displacement of 1.0 mL of water. (Investigation 4)

3. Determine from Graph 2-4 the mass of 1.0 mL of salt solution.

4. Compare the mass of 1.0 mL of salt solution to the mass buoyed by the displacement of 1.0 mL of salt solution. (Investigation 4)

5. Explain any relationship that exists between a liquid's density and its buoyancy.

Report Sheet

Name: _____

Experiment 2

Date: _____ Section: _____

1. Using your best graphing skills, reproduce Graph 2-3 here. Be sure to clearly label each axis and use a ruler to make straight lines.

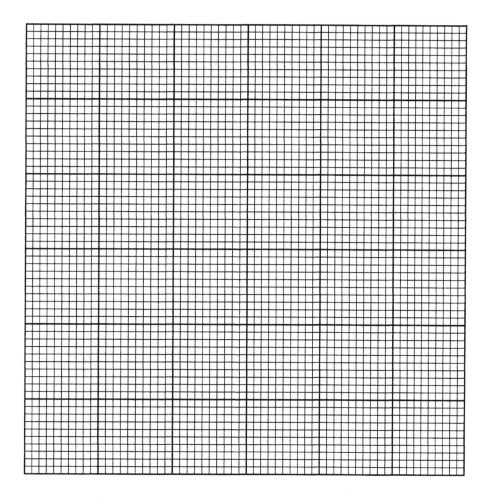

2. Select a convenient point on the above curve and determine both the volume of water displaced and the total mass that was buoyed for that point. Calculate the density of water at that point by dividing the mass buoyed by the displaced volume. Would your answer be different if you selected another point on the curve? Explain your answer.

3. Using question 2 as a guide, calculate the density of the salt water solution.

4. Using the values determined in questions 2 and 3, calculate the specific gravity of the salt water solution.

5. Write a hypothesis that explains the relationship between the volume of a liquid displaced and the mass being buoyed.

6. In your own words, explain how a hydrometer works.

7. Why is a loaded boat more likely to sink in fresh water than salt water? Explain your answer.

Extra Credit

1. A man decided to remove a large rock from his property and put it in a nearby lake. He rolled the rock to the lake and put it in his rowboat. After rowing to the middle of the lake, he pushed the rock overboard. Did the level of water in the lake go up, down, or stay the same? Explain your answer.

Why is the meteorologist never exactly sure of tomorrow's weather?

Uncertainty, Error Bars, and Calibration

Background Concepts

Uncertainties

Every measurement you make contains some degree of uncertainty. For example, suppose you wanted to set up a cannon to deliver pizza from New York to London. How precisely would you have to set the cannon's angle so that the pizza would accurately land at the right person's picnic table? If your measurement were wrong by even 1/1000 of a degree, you might not only miss the table, but you might miss London completely.

So, if your boss asks you what angle the cannon is set to, it's important for you to tell him not just that it's at 30.05 degrees, but also that your measurement might be wrong by ± 0.01 degrees. He might decide not to risk it and end up hitting Belgium.

That ± 0.01 degrees is your uncertainty. Being uncertain doesn't mean that you're incompetent. The uncertainty might be caused by a lot of different things. Maybe the lines on your protractor are too thick. Maybe the ground isn't perfectly level, or perfectly smooth. Maybe the cannon is shaking. Or, maybe the cannon itself isn't perfectly straight, so that the angle itself doesn't have a single, true value, but reads differently at different places along the cannon. Some of these factors are part of the measurement process, and some of them have more to do with the nature of what you're measuring.

Because of all this, whenever we make a measurement, it is best to include an estimate of how far off that measurement might be if it were repeated. So, if I report a measurement of 30 ± 1 degrees, it means that if anybody else repeated the whole measurement, most of them should get between 29 and 31 degrees. It means that if my boss came back saying "Oh yeah? Show me," I should be able to reproduce the same measurement in front of her, give or take 1 degree. A good rule of thumb is that 7 out of 10 people repeating my measurement should get a number that is within the ± range I report.

Functional Graphs

In science, graphs are used to give us a visual image of how things are related to each other. When we vary a parameter on the horizontal axis, how does it change what's on the vertical axis? We could list numbers in a table, saying "When I set the cannon's angle to 10 degrees, I get a range of 1 mile for my projectile; when I set it to 20 degrees …" But a graph makes the trend much more apparent.

Another way in which a graph is useful, is that it allows us to estimate values for cases that we haven't tried. If you measured the range for a setting of 10 degrees and for 20 degrees, you can also see from the graph what the range would likely be for 15 degrees. That's called interpolation. On the other hand, if you only took measurements up to 50 degrees, maybe you could use the graph to project beyond that range of angles and see what it ought to be for a setting of 65 degrees. When you go beyond your data like that, it's called an extrapolation.

Error Bars

Of course, reading between data points or reading beyond them on your graph might depend a lot on how well each data point can be trusted. If each data point has a ± uncertainty, that might make the trend of the points on the graph open to interpretation. That's fine. Discussing different interpretations is part of science too. Above all else, though, we must present our results or graphs as honestly as we can.

To honestly show that there is some ± uncertainty in the data points we put

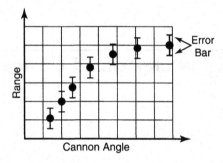

Figure 3–1: Error bars.

on our graph, as shown in Figure 3-1, we draw an error bar around each point. The size of the error bar shows the ± range that corresponds to our ± uncertainty for that measurement.

Investigation 1

Objective

To determine the ± uncertainty for a specific measurement and to determine the uncertainty in a quantity derived from that measurement.

Procedure

1. Set up a ramp system as shown in Figure 3-2, so that you can roll a ball down from the top of the ramp, onto the lab jack. The ball should come onto the lab jack smoothly, without jumping. The lab jack should be set to a height of about 10 – 15 cm:

    actual height: h = _____ cm

2. Mark a starting point at the top of the ramp, and let the ball roll down from there.

3. Tape a blank sheet of paper to the table where the ball tends to land. Draw a line on the sheet where the edge of the lab jack is. If necessary, tape on more sheets (end-to-end) until the paper reaches the edge of the lab jack.

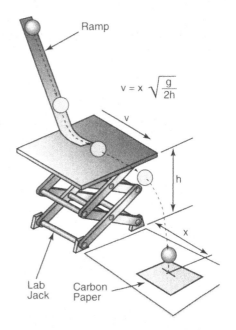

$$v = x \sqrt{\frac{g}{2h}}$$

Figure 3–2: Ramp system.

4. Place a piece of carbon paper over the white paper, where the ball tends to land.

5. Release the ball from its starting point 10 times, so that it makes 10 marks on the white paper.

Interpretation

1. Looking at the spread of dots on your sheet, draw a large "X" to show where the central dot is. This should be the spot that got hit the most. How far is it from the lab jack?

    $$x_{center} = _____ \text{ cm}$$

2. How far is the dot that's furthest from the edge of the lab jack?

    $$x_{max} = _____ \text{ cm}$$

3. How far is the dot that's closest to the edge?

$$x_{min} = \underline{\hspace{1.5cm}} \text{ cm}$$

4. Based on these values, what is the ± range that should be reported?

$$x = x_{center} \pm \underline{\hspace{1.5cm}} \text{ cm}$$

$$x = \underline{\hspace{1cm}} \pm \underline{\hspace{1.5cm}} \text{ cm}$$

5. Suppose you had to roll the ball once more, and predict ahead of time exactly where it would land.

   a. Do you think it could land outside of this range? Explain.

   b. What do you think the chances are that it would land exactly at x_{center}?

   c. What do you think the chances are that it would land within the inside half of the range you wrote down for question 4?

6. Based on the landing position of the ball, you can calculate how fast it was moving when it went across the lab jack:

$$v = x\sqrt{\frac{g}{2h}} = \underline{\hspace{1cm}}\text{meters}\sqrt{\frac{9.8 \text{ m/s}^2}{2 \times \underline{\hspace{1cm}} \text{ meters}}} = \underline{\hspace{1.5cm}} \text{ m/s}$$

7. Repeat this calculation using your values of x_{max} and x_{min} to calculate how much your velocity might be different:

$$v_{max} = \underline{\hspace{1.5cm}} \text{ m/s}$$

$$v_{min} = \underline{\hspace{1.5cm}} \text{ m/s}$$

8. Express the velocity of the ball using a ± range:

$$v = \underline{\hspace{1.5cm}} \pm \underline{\hspace{1.5cm}} \text{ m/s}$$

Investigation 2

Objective

To plot a graph of a functional relationship.

To reduce the data to a linear form and plot it as a straight line.

To extrapolate and interpolate data from a graph.

To plot error bars on the data graphed, according to the uncertainties in measurements.

Procedure

1. Hang a small weight as shown in Figure 3-3. Adjust the string length so the weight hangs down 30 cm.

2. Using a stop watch, measure how long it takes for the weight to swing back and forth 10 times.

3. Record your result in Table 3-1 below. Shorten the string and repeat your measurements for all the lengths listed in Table 3-1.

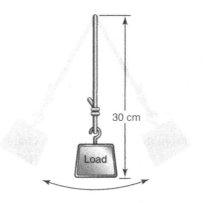

Figure 3–3: Weight on a string.

Hanging Distance	Time for 10 Swings
30 cm	seconds
20 cm	seconds
10 cm	seconds
5 cm	seconds
2 cm	seconds

Table 3–1: Time for 10 swings.

4. Now plot this data as points on the Graph 3-1. Draw a *smooth* curve through the points. The curve should pass close to each point, but it doesn't have to exactly touch each point.

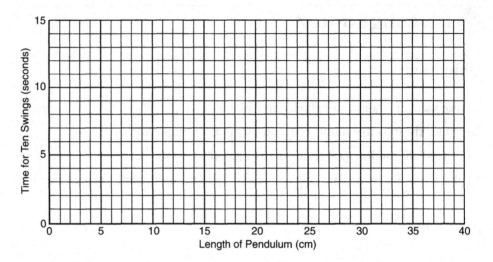

Graph 3–1: Pendulum swing data.

Interpretation

Interpolation

1. From the line you drew on Graph 3-1, determine how long the string would have to be for the weight to swing 10 times in 5 seconds.

2. How long would it have to be to swing 10 times in 4 seconds?

3. How much time would it take for a pendulum that was 15 cm long to swing 10 times?

Extrapolation

4. Based on the data from Graph 3-1, approximately how much time would it take for a pendulum that was 40 cm long to swing 10 times?

5. Theoretically, a pendulum that's infinitesimally short (length is approximately zero), should take no time at all to swing back and forth. Does Graph 3-1 agree with this? Explain.

Functional Relationship

6. Let's see if we can make all this easier. Complete Table 3-2. Plot your data again on Graph 3-2, but this time use the square of each of the time values.

Hanging Distance	Time 10 swings	Time-squared 10 swings
30 cm	seconds	seconds²
20 cm	seconds	seconds²
10 cm	seconds	seconds²
5 cm	seconds	seconds²
2 cm	seconds	seconds²

Table 3–2: Pendulum data squared.

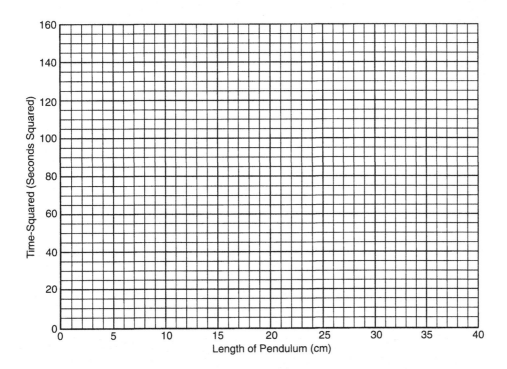

Graph 3–2: Pendulum data squared.

7. Your points should form a straight line on Graph 3-2. Draw a straight line that passes through your data points, or as close to each point as possible.

8. Based on this line, what should be the value of time-squared for a pendulum length of 40 cm?

---

9. How is it easier to extrapolate and interpolate data points in this form?

10. So far we have not mentioned uncertainty. Suppose now that each data point you plotted on Graph 3-2 had an uncertainty of ± 5 s^2. We would have to represent that by drawing "error bars" around each point as shown in the error bar section. Draw error bars on each point corresponding to ± 5 s^2.

11. Draw 2 other straight lines that pass through the ends of as many error bars as possible. These 2 lines should be as different from each other as possible.

12. Based on these lines, what would the values of the time-squared for a pendulum of 40 cm? (One for each line).

13. Use this to estimate what the $\pm$ uncertainty would be in this extrapolated value.

Investigation 3

Objective

To draw a calibration curve for a measuring instrument and use the calibration curve to adjust measurements to agree with accepted standards.

Procedure

1. You will be provided with a digital multimeter, and three standard resistors. The multimeter as you receive it does not measure resistor values properly according to I.P.L.* standards. To correct for this, however, the three resistors you are given have been labeled with their correct I.P.L. resistor values, for you to use in calibrating your multimeter.

2. Set your multimeter to the 100 Ω (ohm) setting. By touching both sides of each resistor with the multimeter's probes, record what your multimeter reads for each of the standard resistors in Table 3-3.

I.P.L. Standard	Measured Value
10 ohms	ohms
20 ohms	ohms
50 ohms	ohms
30 ohms	ohms

Table 3–3: Measured values of resistors.

3. Connect the 10 and 20 ohm Standard resistors together as shown in Figure 3-4. The combined I.P.L. Standard value is 30 ohms. Measure what the combination reads with your multimeter, and record what your multimeter reads in Table 3-3.

4. For each of the values in Table 3-3, plot a data point on Graph 3-3. The graph plots what your multimeter reads as a function of what the I.PL. Standard value is.

5. Based on your 4 points, draw a line that represents the behavior of this graph. You may use a straight or curved line, at your own discretion.

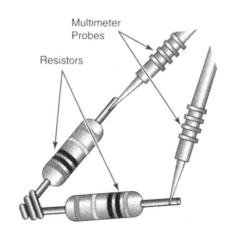

Figure 3–4: Connect and test resistors.

* I.P.L. is the "I'm Pulling your Leg" standard. The standard resistors you were provided may not be exactly true to their labeled values.

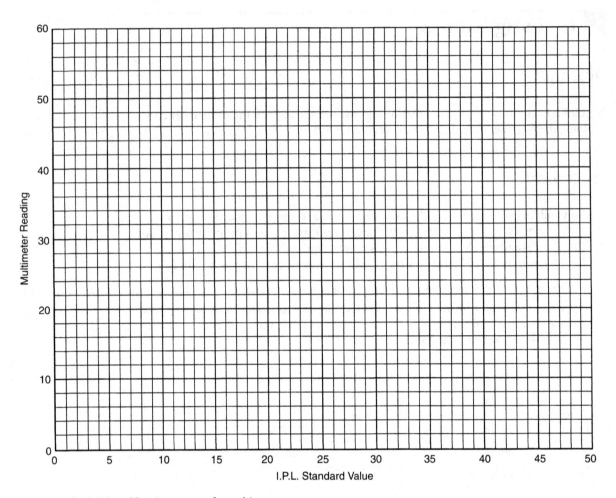

Graph 3–3: I.P.L. calibration curve of a multimeter.

6. Every team in class will be given a single unlabeled resistor to measure. Each team will have 60 seconds to make a measurement of this "mystery" resistor with the multimeter. Because the multimeter doesn't match the I.P.L. standard exactly, you must adjust your measurement. According to the line on your graph, what is the I.P.L. standard value of the "mystery" resistor that would give you the same reading as the multimeter? Specify 2 decimal places.

mystery value = _____ I.P.L. standard ohms

7. Each group in the class will come up with their own value. Estimate the variation you expect to see in the final I.P.L. standard values they come up with:

All the groups' values will probably be within:

± _____ I.P.L. standard ohms of each other

SCORING: You will be given points for how close your value in step 5 comes to the average of the class' results. You will also be given points for arriving at a ± deviation that includes almost all of the class' results, while being narrow enough not to include the extreme highest and lowest results.

8. In Table 3-4, list the results obtained by all the lab groups, along with their ± uncertainties. (Remember to include your own.)

Group Name	Measurement in I.P.L. Standard Ohms (with ± uncertainty)

Table 3–4: Lab group measurements.

average value = _____ I.P.L standard ohms

9. Add up your score for this activity:

  a. Were you within:

    1 ohm of the average? + 1 point

    0.5 ohms of the average? + 2 points

    0.2 ohms of the average? + 3 points = _____

  b. Is the average result within the ± uncertainty range you estimated? + 2 points = _____

  c. Is your ± uncertainty range big enough? How many of the other groups' results are within the ± uncertainty range you estimated?

    2–3 of them… + 1 point

    at least half of them + 3 points = _____

  d. Is your ± uncertainty range too large:

    Does it exclude the largest result? + 1 point

    Does it exclude the smallest result? + 1 point

    Does it exclude both? + 3 points

    (Let's figure that these guys are just a little too far off) = _____

                    TOTAL: = _____

Interpretation

1. Why would your measurement be more accurate if the value of the unknown resistor were close to one of the standard resistors?

2. Suppose one of the standard resistors you were given had been labeled incorrectly. How would that affect your graph, and your result for the mystery resistor?

3. How do you suppose the values of the I.P.L. standard resistors were established?

Investigation 4

Objective

To calibrate a pH meter.

Procedure

Some measuring devices do most of the calibration work for you. One of them is a pH meter. Still, you must provide it with the standard solutions that the instrument uses to calibrate itself.

1. Obtain a digital pH meter from your instructor.

2. Obtain 3 standard buffer solutions from your instructor: pH 4 and pH 12, along with some deionized (distilled) water.

3. Press the calibration button for "pH 7" on the pH meter and immerse the probe into the distilled water. Depending on your meter, you will probably need to adjust it to read "7."

4. Dry the probe with a tissue. Press the calibration button for "pH 4" and immerse the probe into the pH 4 standard solution. Depending on your meter, you might need to adjust it to read "4."

5. Rinse the probe with deionized water and again dry with a tissue. Press the calibration button for "pH 12" on the pH probe and insert it into the pH 12 standard solution. Depending on your meter, you will probably need to adjust it to read "12."

6. Your meter should now be calibrated. Your instructor will provide you with a "mystery solution." Pour off a small sample of the solution to measure its pH. Be sure to rinse the probe with deionized water and dry it with a tissue before using it to determine the pH of the "mystery solution."

mystery solution pH = _____

Interpretation

1. Collect the results for the pH of the "mystery solution" from all other groups.

2. What is the average value of the pH?

3. What is the ± variation in the groups' results?

4. Suggest what may have caused the deviation in results.

5. Why is it generally a good idea for the pH meter to be calibrated using more than one standard solution?

Report Sheet

Experiment 3

Name: _____

Date: _____ Section: _____

1. Report your conclusions from Investigation 1 regarding where the ball would most likely land if you rolled it again:

   a. Ball's landing spot: _____ ± _____ cm from track

   b. With a horizontal velocity of: _____ ± _____ m/s

2. Using your best graphing skill, reproduce the graph in Investigation 2, Time for 10 Swings-squared vs. pendulum length. Include the error bars and fit lines.

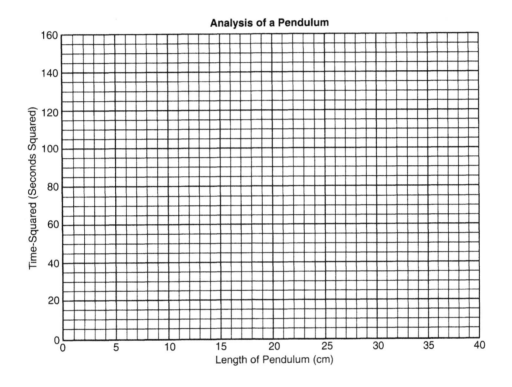

Analysis of a Pendulum

3. What is the extrapolated value of time-squared for a pendulum that's 40 cm long?

   _____ ± _____ s^2

4. Report your estimate of the I.P.L. standard resistance of the mystery resistor in Investigation 3.

   _____ ± _____ ohms

   class average: _____

   TOTAL POINTS from score sheet: _____

5. If 50 people measure the same thing using similar, but not identical, equipment why don't they all get the same result?

6. Why can't you say who has the "true" value?

7. Meteorologists measure barometric pressures, temperatures, and wind currents at specific places, but not everywhere. What effect do measurement uncertainties have on their forecasts?

Extra Credit

1. Discuss whether you agree with the following:

"There is no truth, except what we measure. Since our measurement of anything, say a resistor, must have some uncertainty, we can never know for sure what the true value is. Therefore, the resistor doesn't really have a 'true' value."

What makes a popcorn kernel pop?

Percent Composition and Error Analysis

Background Concepts

Most foods contain water. In items such as milk, soft drinks, fruit juices, watermelon, etc., the water is not only obvious, but it is also responsible for most of the mass. In these examples, we recognize the water because it is in its familiar liquid state. Water is less obvious in baked items and meats, but we have all experienced the change in texture that occurs when pastries dry out or meat becomes jerky. Some foods like coffee, peanuts, popcorn kernels, and dried beans seem to contain no water at all. Most likely our reasoning for this conclusion is because they do not feel moist.

You may recall that water is a polar molecule. This means that water molecules have sites that are either partially negative ($\delta-$) or partially positive ($\delta+$) and consequently they tend to cling to each another. When a sufficient number of water molecules are present, they cluster together forming liquid water. This same polar nature, however, also causes water molecules to be attracted to other polar and ionic substances. Water molecules, therefore, tend to chemically bind with these substances forming hydrated materials. In hydrated materials, the water is not in the liquid form and, therefore, the substance does not feel moist to the touch. Some hydrated substances, such as hydrated salts, tend to contain a specific amount of water. Therefore, chemical formulas such as $CuSO_4 \cdot 5H_2O$ or $BaCl_2 \cdot 2\,H_2O$ can be used to represent them. The dot between the two parts of the formula is used to indicate that the water is still a separate molecule, but it is chemically bound to the ions of the salt. Many hydrated substances, such as pastries or meats, contain variable amounts of water that can only be represented by a percent of its composition.

When a sufficient amount of energy is added to a hydrated material, a chemical reaction occurs in which the bound water is released. The following equation represents the dehydration of the hydrated salt known as barium chloride dihydrate:

$$BaCl_2 \cdot 2\,H_2O \;\rightarrow\; BaCl_2 + 2\,H_2O$$

$$\text{hydrate} \;\rightarrow\; \text{anhydrous salt} + \text{water}$$

From the equation, it is apparent that the heating process does not produce any new substances, but merely separates the bound water from the anhydrous salt. By comparing the mass of the hydrated salt and the mass of anhydrous salt remaining after heating, it is possible to calculate the percentage of water in the hydrate. The following formula is general and can be used to calculate the percent of water in any material containing water of hydration:

$$\% \text{ water} = \frac{\text{mass lost by heating}}{\text{mass before heating}} \times 100$$

To solve the above equation, it is obvious that several mass measurements are required. In a science laboratory, the instrument used for measuring mass is the balance. Balances come in a variety of sizes and shapes. The one thing all balances have in common is that the object being weighed is loaded on one end of a beam supported by a fulcrum and the counterbalancing masses are loaded on the other end. In this way, balances are very different from the familiar bathroom scales where your weight is determined by the amount of work done in winding a spring. Regardless of the type of balance, scale, or other instrument used, all measurements are subject to errors in both precision and accuracy.

Precision has to do with the reproducibility of results using the instrument. As a rule of thumb, the ability of an instrument to give reproducible measurements can often be gauged by its cost – the higher the cost, the higher the precision. Accuracy has to do with the closeness of a measurement to the true or accepted value. For example, if many arrows are shot at a target and they all land in a small circle away from the bull's-eye, we would have precision, but not accuracy. On the other hand, if they all land in a small circle around the bull's-eye, we would have both precision and accuracy – the goal for all instruments.

Since we do not live in a perfect world, what can we do to improve the quality of data that is collected and reported from instruments? First, always use high-quality instruments that provide the degree of precision and accuracy necessary for the task. Second, learn how to use the instruments properly. Even then, errors will occur. Errors in making experimental measurements tend to fall into two classes: systematic errors and random errors.

Systematic errors are correctable because they are caused by a consistent factor such as having the end and the number scale of a ruler start at different points. Once the difference is known, a constant factor can either be added or subtracted from all measurements to correct for the error. Systematic errors cannot be eliminated by merely repeating the measurement, since the same error is involved in each measurement. Random errors, on the other hand, can be detected by fluctuations in repeated measurements. These variations are beyond the control of the observer. After making a series of readings, it will be found that they fluctuate around a mean (average) value.

Manufacturers of analytical balances report a precision of ± 0.0001 g, compared to ± 0.01 g for electronic top-loading balances. This type of valuable information can be included as a part of an object's mass by writing 34.56 ± 0.01 g. This alerts the reader that the mass of the object could be as low as 34.55 g or as high as 34.57 g. If the mass of this object were determined repeatedly, then it would be possible to calculate its mean value.

Trial #1	34.56 ± 0.01 g
Trial #2	34.58 ± 0.01 g
Trial #3	34.57 ± 0.01 g
Trail #4	34.55 ± 0.01 g
Trail #5	34.57 ± 0.01 g

The mean value is determined by adding the separate measured values and dividing the sum by the number of measurements.

$$\text{mean value} = \frac{\text{sum of measurements}}{\text{number of measurements}} = \frac{172.83 \pm 0.05 \text{ g}}{5} = 34.57 \pm 0.01 \text{ g}$$

This value, 34.57 ± 0.01 g, should be considered as closer to the true value than any randomly chosen single measurement. Whenever possible, therefore, use the average of several readings rather than a single measurement.

The next consideration is how these measurement errors influence the results. For example, if two measurements, each with an uncertainty of ± 0.01 g, are required to determine the amount of water lost from a hydrate, what will be the uncertainty in that value? A simple method for calculating the error in the results is to calculate the maximum error that would result if the error in each measured quantity was its maximum value. Although it is unlikely that the errors would combine in this way, this method presents the worst possible case. For example, if the mass of a hydrate was 54.46 ± 0.01 g and after heating it had a mass of 34.89 ± 0.01 g, the mass of the water lost would be:

$$
\begin{array}{r}
54.46 \ \pm 0.01 \text{ g} \\
- 34.89 \ \pm 0.01 \text{ g} \\
\hline
19.57 \ \pm 0.02 \text{ g}
\end{array}
$$

From this example it can be seen that although the mathematical process was subtraction, the ± 0.01 g uncertainty associated with each measurement was added when determining the difference. Simply stated, the maximum error in a sum or difference calculation is the sum of the absolute values for each of the measured quantities.

When the calculated value is determined by multiplication and/or division operations, the problem becomes a bit more complicated. First, the uncertainty in each measurement must be converted into a percent of uncertainty. Second, the percentages of uncertainty are added. For example, suppose you were to determine the density of a liquid and found that 9.8 ± 0.2 mL of it had a mass of 10.54 ± 0.01 g. The calculations would be as follows:

$$\text{density} = \frac{10.54 \pm 0.01 \text{ g}}{9.8 \pm 0.2 \text{ mL}} = \frac{10.54 \text{ g} \pm 0.09\%}{9.8 \text{ mL} \pm 2.04\%}$$

$$= 1.08 \text{ g/mL} \pm 2.13\% \text{ or } 1.08 \pm 0.02 \text{ g/mL}$$

In the experimental sciences, when the true or accepted value is known, it is common practice to calculate the percent error. This is done in much the same way as you would determine the percent wrong on an exam. You divide the number of questions missed by the total number of questions on the test and multiply by 100. For example, if you missed 4 out of 82 questions, then $4/82 \times 100 = 4.9\%$ wrong, but your percent correct would be $78/82 \times 100 = 95.1\%$ correct. As can be seen from this example, the percent correct plus the percent error should always equal 100 percent.

Consider another example. If from experimental data you determine that the percent of water in a hydrated substance is 15.65%, but the accepted value is 13.50%, what would be the percent of error? When calculating percent error, the absolute difference is used for the numerator term and the accepted value for the denominator term. Examine the following:

$$\frac{|\text{accepted value} - \text{experimental value}|}{\text{accepted value}} \times 100 = \% \text{ error}$$

$$\% \text{ error} = \frac{|13.50\% - 15.65\%|}{13.50\%} \times 100 = \frac{2.15}{13.50} \times 100 \cong 15.92\%$$

Regardless of the measurement method or error analysis performed, the single purpose remains to inform the reader of everything known about the value – nothing more and nothing less.

Investigation 1

Objective

To explore the behavior of a hydrated substance.

Safety Requirements

- Always wear splash proof safety goggles
- Do not touch chemicals
- Do not point test tubes being heated toward another person.
- Wash skin that is exposed to chemicals in cold water. Then, notify the teacher.
- Dispose of reagents according to directions

Procedure

1. Select and determine the mass of a 10 mL test tube.

    ———— g

2. Place a small quantity of $CoCl_2 \cdot 6 H_2O$, approximately the size of a green pea, into the weighed test tube. Observe and record its appearance.

3. Return to the same balance as used in step 1 and determine the mass of the test tube and hydrate. Record the combined mass.

    ———— g

4. Using a test tube clamp to hold the tube at an angle, gently heat the tube's contents. Slowly move the tube back and fourth through the burner flame. Watch the cooler, open end of the test tube for any evidence that water is being removed from the hydrate. Move the flame toward the open end of the tube and continue heating only until the color change is complete. Record your observations.

5. Allow the tube to cool briefly by standing it in a small (150 mL) beaker.

6. Once the tube has reached room temperature, return to the same bal-

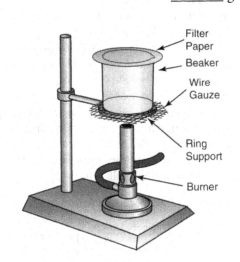

Figure 4–1: Filter paper drying set-up.

ance as previously used and again determine the mass of the tube and its contents. Record the combined mass.

_____ g

7. Touch the bottom of the test tube. Using a dropping pipette, add 3 drops of DI water to the contents of the tube. Record your observations.

8. Using the dropping pipette, add enough more water to the test tube to dissolve the remaining solid. Place a piece of filter paper over a 150 mL beaker and wet it with this solution. As shown in Figure 4-1, place the beaker and moistened filter paper on a wire gauze or on a hot plate and warm it gently, until the filter paper is dry. Do not allow the paper to catch on fire. Record your observations.

9. Using a dropping pipette, place one drop of water on a colored section of the filter paper. Record your observations. Save the remainder of the paper for use in Investigation 2.

Interpretations

1. What is the color of

   a. Hydrated $CoCl_2 \cdot 6\ H_2O$?   _____

   b. Anhydrous $CoCl_2$?   _____

2. Did the hydrate gain or lose mass when it became the anhydrous salt?

3. Based on your measurements, what is the percent of water in the hydrate? Show your calculations below.

4. When a hydrated salt is converted to an anhydrous salt, is the process endothermic (heat in) or exothermic (heat out)? Write an equation for the process, placing the term "energy" on the correct side of the arrow.

5. In your own words, explain how a piece of the dry, colored paper you made in step 8 could be used by a manufacturer to determine if the seal had been broken on a package of vitamin tablets.

■ Investigation 2

Objective

To develop a method for determining the percent of water in hydrated substances.

Procedure

1. Obtain a cardboard cover and wire stirrer to use with your 150 mL beaker. Determine their combined mass by repeatedly weighing them on the same balance to the nearest ± 0.01 g. After completing and recording the first mass determination, remove the beaker, cover and stirrer, re-zero the balance and then reweigh. Repeat this process a third time, recording each of the three separate values in Table 4-1.

2. Place between 12 and 15 kernels of popcorn in the beaker. Record the actual number used in Table 4-1.

3. Return to the same balance as used in step 1, and determine the combined mass of the beaker, cover, wire, and popcorn. Make three separate determinations and record the values in Table 4-1.

4. Place the beaker, cover, wire, and popcorn on a wire gauze supported by a ring stand, as shown in Figure 4-2. Put a rolled piece of the blue cobalt(II) chloride test paper into the beaker spout and under the cardboard cover. Move the burner flame back and forth across the bottom of the beaker, while moving the kernels with the wire stirrer. After the first three or four kernels pop, stop heating with the burner. The residual heat is enough to get the rest of the kernels to pop without scorching any of the popped kernels. Record any observations that suggest that water is a product of the popping process.

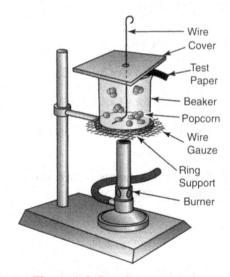

Figure 4–2: Popping corn set-up.

5. Allow the beaker, cover, wire, and popped corn to cool to room temperature. Determine the number of kernels that did not pop. Record the number in Table 4-1.

6. Return to the same balance as used in steps 1 and 3, and again determine the combined mass of the beaker and its contents. Make three separate mass determinations and record them in Table 4-1.

Item	Trial No.		Calculated Mean Values
Beaker/Cover/Wire	#1	g	
	#2	g	
	#3	g	
Beaker/Cover/Wire & Kernels	#1	g	
	#2	g	
	#3	g	
Beaker/Cover/Wire & Contents	#1	g	
	#2	g	
	#3	g	
Number of Kernels Used			
Number of Unpopped Kernels			
Mass of All Kernels			
Average Mass of Each Kernel			
Number of Kernels Popped			
Mass of Kernels Popped			
Mass Lost in Popping Process			
Percent of Water in Kernels that Popped			
Average Number of Grams of Water in a Popcorn Kernel			

Table 4–1: Investigation 2 data and calculations.

Interpretation

1. Using the method discussed in the Background Concepts section, calculate the mean value for the combined masses of the beaker, cover, and wire stirrer. Record the value in Table 4-1.

2. Using the same method as in step 1 above, calculate the mean value for the combined masses of the beaker, cover, wire stirrer, and popcorn kernels. Record the value in Table 4-1.

3. Using the same method as in steps 1 and 2, calculate the mean value for the combined masses of the beaker, cover, wire stirrer, and contents after heating. Record the value in Table 4-1.

4. Using the mean values, calculate the mass of the popcorn kernels used. Show your work below and record the value in Table 4-1.

5. Calculate the average mass of a popcorn kernel. Show your work below and record the value in Table 4-1.

6. Calculate the number of kernels of corn that popped and record it in Table 4-1.

7. Using the average mass of a popcorn kernel (step 5), calculate the total mass of the kernels that popped. Show your work below and record the value in Table 4-1.

8. Using the mean values, calculate the mass loss that occurred as a result of the heating process. Show your work below and record the value in Table 4-1.

9. What evidence is there that loss of water is responsible for the mass loss?

10. Using the values calculated in steps 7 and 8, calculate the percent of water in the kernels of corn that popped. Refer to the Background Concepts section if you need help with the calculation. Show your work below and record the value in Table 4-1.

11. In the space below, describe the three measurements that are necessary to calculate the percent of water loss for any substance.

12. Using the values calculated in steps 5 and 8, determine how much water is in an unpopped kernel of corn. Show your work below and record the value in Table 4-1.

13. Based on your experimental findings, how would you answer the question, "What makes a kernel of popcorn pop? Write your explanation in the space below.

Investigation 3

Objective

To calculate the percent error in an experimentally determined value.

Procedure

1. Obtain the data for completing Table 4-2 by transferring it from Table 4-1 and the Investigation 1 results from four other students.

2. Using the other student's results, calculate the mean value for each of the items. Show your work below and record the values in Table 4-2.

Item	Your Result	Student				Calculate Mean Value	Percent Error
		1	**2**	**3**	**4**		
Average Mass of Each Kernel							
Percent of Water in Popcorn Sample							
Average Number of Grams of Water in a Popcorn Kernel							

Table 4–2: Calculated student results.

Interpretation

> **Note:** The mean values for the other students' data are to be used as the accepted values when doing the following calculations.

1. Calculate the percent error in your average mass of each kernel determination, using the mean value as the accepted value. Show your work below and record the results Table 4-2.

2. Calculate the percent error in your value for the percent of water in the popcorn sample, using the mean value as the accepted value. Show your work below and record the results Table 4-2.

3. Calculate the percent error in your value for the number of grams of water in a kernel of popcorn, using the mean value as the accepted value. Show your work below and record the results Table 4-2.

4. The average error, a, is the average of the deviations from the mean or accepted value. The formula for its calculation is:

$$a = \pm\frac{\Sigma(d)}{n}$$

where Σ means the sum of the individual deviations, d, divided by the number of measurements, n. The plus or minus, $\pm$, signs means it makes no difference if the measurement is above or below the mean value, they are still added together.

5. Use the mean value for the number of grams of water in a kernel of popcorn as the accepted value. Determine the deviation of your data and each of the other four student's data from this value. Find the sum of the various deviations and divide it by the number of measurements. Complete Table 4-3 by determining the average error for the number of grams of water in a popcorn kernel.

Student	Average Number of Grams of Water in a Popcorn Kernel	Accepted Value from Table 4-2	Calculated Deviation d	Sum of Deviation Σ	Average Error a
Your Data					
1					
2					
3					
4					

Table 4–3: Average error in the number of grams of water in a popcorn kernel.

6. Examine the calculated deviations, d, in Table 4-3. Is there one that appears to be much larger than all the others? Would there ever be a situation where one might consider "throwing away data" because it appears to contain a serious error? Explain your answer.

Report Sheet

Experiment 4

1. What is the color of:

    a. Hydrated $CoCl_2 \cdot 6\ H_2O$?   _____

    b. Anhydrous $CoCl_2$?   _____

2. Explain the role energy plays in the conversion of a hydrate to its anhydrous salt.

3. Transfer your data and calculated values from Table 4-1 to Table 4-4.

Item	Trial No.		Calculated Mean Values
Beaker/Cover/Wire	#1	g	
	#2	g	
	#3	g	
Beaker/Cover/Wire & Kernels	#1	g	
	#2	g	
	#3	g	
Beaker/Cover/Wire & Contents	#1	g	
	#2	g	
	#3	g	
Number of Kernels Used			
Number of Unpopped Kernels			
Mass of All Kernels			
Average Mass of Each Kernel			
Number of Kernels Popped			
Mass of Kernels Popped			
Mass Lost in Popping Process			
Percent of Water in Kernels that Popped			
Average Number of Grams of Water in a Popcorn Kernel			

Table 4–4: Investigation 2 data and calculations.

4. Based on your experimental results, what percent by weight, of an unpopped kernel is water? Show your work below.

5. Based on your experimental results, if you were to purchase a 1 pound bag of unpopped corn, how many grams of water would be included in its weight? Show your work below.

6. From Investigation 3, what was the average error, a, in your calculated number of grams of water in a kernel of popcorn?

_____ g

7. What role does water play in causing a kernel of popcorn to pop? Explain your answer.

Extra Credit

1. In the Interpretation section of Investigation 2, you were asked to use the values calculated in steps 7 and 8 to determine the percent of water in the popped corn sample. Repeat the calculation in the space below, but convert the uncertainty in each mass measurement into its corresponding percent and determine the percent error in the calculated value.

2. Another common statistical method for analyzing data is through the calculation of the standard deviation, δ. The standard deviation is a statistic that tells you how closely the data is clustered around the mean. It is calculated by taking the square root of the average of the squares of the deviations from the mean. The formula is:

$$\delta = \pm\sqrt{\frac{\Sigma(d)^2}{n-1}}$$

In general, if most of the data is very close to the mean, then the standard deviation (δ) value will be small. If, however, the data is widely scattered from the mean, then the standard deviation will be large.

Using Interpretation step 4 in Investigation 3 as a model, calculate the standard deviation for the average number of grams of water in an unpopped kernel of corn. Show your work below and comment on what the calculated value means.

How can you determine the atomic mass of a metal?

Estimating the Atomic Mass of Metals

■ Background Concepts

The number of sheets of paper in a large stack can be easily determined from its total mass, if you know the mass of a single sheet. Determining the number of atoms present is a similar problem, but to do so you must know the mass of a single atom. For many years, finding a way to determine the atomic mass of an element was the missing link to determining the ratio of elements in a compound.

Since the infinitesimal size of an atom makes direct measurement impossible, chemists sought other measurable properties that could be related to the atomic mass. The first successful method was reported in 1819 by two French chemists Pierre-Louis Dulong (1785–1838) and Alexis-Thérèse Petit (1791–1820). Through experimentation, they determined that there was a nearly constant relationship between the specific heat and the atomic mass of solid elements; the specific heat being defined as the amount of energy required to change the temperature of 1 gram of the substance by 1°C. They stated, "The atoms of all simple bodies have exactly the same capacity for heat." Today, the law of Dulong and Petit states that the product of the specific heat and molar mass for all solid elements is approximately constant. Subsequent investigators demonstrated that this relationship had severe limitations, but it is still considered a useful relationship for determining the approximate atomic mass of the heavier metallic elements.

In this experiment, you will develop a method for determining the specific heat of known metals. By pooling data with your classmates, you will be able to derive a mathematical relationship that will allow you to determine the approximate atomic mass of an unknown metal.

Investigation 1

Objective

To determine the equilibrium temperature and the amount of heat transferred by mixing hot and cold water samples.

> ### Safety Requirements
>
> ■ Always wear splash proof safety goggles.
>
> ■ Do not touch hot objects.
>
> ■ Use beaker tongs or towels when handling hot glassware.
>
> ■ Return used metal to the container designated by your instructor.

Procedure

> **NOTE:** Your instructor may direct you to work with a partner in recording the time-temperature data. All data collected in this investigation should be recorded in Table 5-1.

1. Using a hot plate or ring stand, ring, wire gauze, and burner, set up the equipment as shown in Figure 5-1.

2. Pour 100.0 g of DI water into a 400 mL beaker. (In this experiment, you may assume that 1.00 g H_2O = 1.00 mL H_2O.) Cover the beaker with a watch glass to reduce evaporation, and heat to a temperature of 45–55°C.

3. Pour a second 100.0 g of room temperature DI water into your calorimeter cup. Read and record its temperature in Table 5-1.

4. Record the temperature of the hot water in Table 5-1, then rapidly pour it into the cold water. The time of mixing is called 0 seconds.

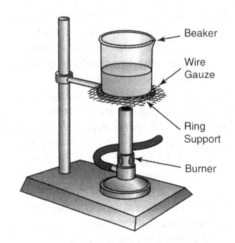

Figure 5–1: Ring stand set up.

5. Record the temperature of the mixed sample every 30 seconds. The first temperature should be recorded 30 seconds after mixing. Continue the readings for the next 3–5 minutes. When not actually reading the thermometer, stir the mixture.

6. Plot the time-temperature data recorded in Table 5-1 on Graph 5-1 and draw a smooth curve among the points.

Mass of Cold Water		g	Temperature of Cold Water		°C
Mass of Hot Water		g	**Temperature of Hot Water**		°C
Lapse Time After Mixing, s	**Temperature, °C**		**Lapse Time After Mixing, s**	**Temperature, °C**	
0			210		
30			240		
60			270		
90			300		
120			330		
150			360		
180			390		
Predicted Highest Temperature From Extrapolation				°C	

Table 5–1: Hot and cold water mixing data.

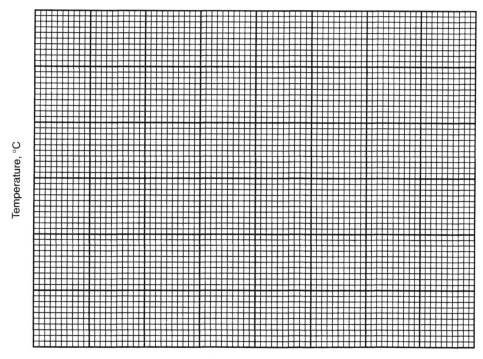

Time, s

Graph 5–1: Time-temperature plot for hot and cold water.

Interpretation

1. Examine the slope of your curve starting 60 seconds after mixing the hot and cold water. What happens to the temperature with increasing time?

2. What does the time-temperature curve indicate about the transfer of heat between the calorimeter and its surroundings? Explain.

3. At what time, with respect to mixing, did transfer of heat to the surroundings start?

4. As shown in Figure 5-2, by means of a straight edge extrapolate the curve of the graph to the time of mixing or zero seconds. Read and record this predicted temperature in Table 5-1.

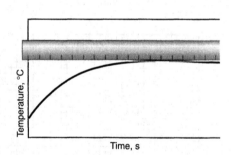

Figure 5–2: Extrapolation of cooling curve to time zero using a straight edge.

5. Assume that your calorimeter allowed no heat transfer to its surroundings. What does the temperature recorded in question 4 represent? Explain.

6. Using the predicted temperature as the equilibrium temperature for the hot and cold water, calculate and record in Table 5-2 the temperature change, ΔT, for the hot and the cold water.

	Initial H_2O Temperature, °C	Equilibrium Temperature, °C	Temperature Change, ΔT
Hot H_2O			
Cold H_2O			

Table 5–2: Temperature changes for hot and cold water.

7. Using 4.186 J/g · °C for the specific heat (sp ht) of water calculate and record in Table 5-3 the amount of heat lost by the hot water and the amount of heat gained by the cold water. Remember: Heat = sp ht $\times g \times \Delta T$.

	Heat Gained = sp ht $\times g \times \Delta T$	Heat Lost = sp ht $\times g \times \Delta T$
Hot H_2O		
Cold H_2O		

Table 5–3: Heat lost and gained by hot and cold water.

8. How does the heat lost by the hot water compare to the heat gained by the cold water? Explain any similarities or differences.

Investigation 2

Objective

To determine the specific heat of a metal by calorimetry

Procedure

1. Ask the instructor to assign one of the known metals. Record its name in Table 5-4. Weigh out a sample of approximately 40–50 g and record its mass to the nearest ± 0.01 grams in Table 5-4.

2. Place this sample into a clean dry calorimeter.

3. Pour 100.0 g of DI water near room temperature into the calorimeter containing the metal. Stir the contents for 1 minute and record the temperature in Table 5-4.

4. Heat 100.0 g of DI water to a temperature between 45° and 55°C.

5. Record the temperature of the hot water in Table 5-4. Rapidly add the hot water to the contents of the calorimeter. Start timing.

6. Record the temperature of the mixture every 30 seconds. The first temperature should be recorded 30 seconds after mixing. Continue the readings and recording as in Investigation 1.

7. Prepare Graph 5-2 of the time-temperature data recorded in Table 5-4.

8. Extrapolate the cooling portion of the curve to zero seconds. Record in Table 5-4 and use this predicted temperature as the equilibrium temperature for the calculations in Table 5-5.

Name of Metal Used				
Mass of Metal Used	g	**Temperature of Cold Water and Metal**		°C
Mass of Hot Water Used	g			
Mass of Cold Water Used	g	**Temperature of Hot Water**		°C
Lapse Time After Mixing, s	**Temperature °C**	**Lapse Time After Mixing, s**		**Temperature °C**
0		210		
30		240		
60		270		
90		300		
120		330		
150		360		
180		390		
Predicted Highest Temperature From Extrapolation				°C

Table 5–4: Metal and water data.

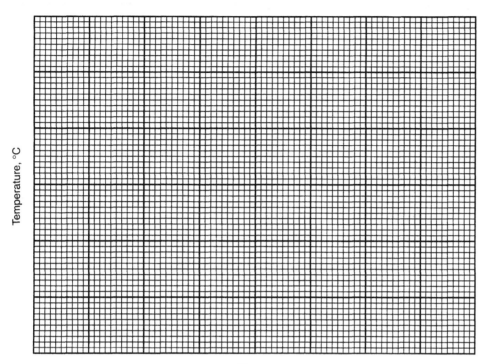

Graph 5–2: Time-temperature plot for water and metal.

Interpretation

1. Using the information recorded in Table 5-4 calculate the specific heat of the metal. Use Table 5-5 as a guide.

	Initial Temperature °C	Equilibrium Temperature °C	Temperature Change ΔT
Hot Water			
Cold Water			
Metal			
Heat Gained		=	Heat Lost
Cold Water	+	Metal =	Hot Water
sp ht$_{water} \times \Delta T_{water} \times g_{water}$	+	sp ht$_{metal} \times \Delta T_{metal} \times g_{metal}$ =	sp ht$_{water} \times \Delta T_{water} \times g_{water}$
		=	
sp ht$_{water}$ = 4.186 J/g · °C		sp ht$_{metal}$ =	J/g · °C

Table 5–5: Calculation of the specific heat of known metal.

2. How do the metal and water specific heats compare? Which is higher?

Investigation 3

Objective

To determine the relationship between the atomic mass and specific heat of a metal.

Procedure

1. Table 5-6 lists the specific heat for several metals taken from the literature. Add your calculated specific heat recorded in Table 5-5.

2. If possible, obtain specific heats from your classmates for three other metals. Record their names, calculated specific heats, and atomic mass from the periodic table to complete Table 5-6.

Data Source	Name of Metal	Atomic Mass of Metal, g/mole	Specific Heat of Metal, J/g · °C
Literature	Antimony	121.8	0.209
Literature	Cadmium	112.4	0.230
Literature	Chromium	52.00	0.460
Literature	Platinum	195.0	0.134
Literature	Silver	107.9	0.239
Your Data			

Table 5–6: Data: Specific heat and atomic mass of each metals used.

3. Using the atomic mass and specific heat data recorded in Table 5-6, prepare Graph 5-3.

4. Draw a straight smooth line among the plotted values.

5. Select four points, other than data points, along the line produced on Graph 5-3. Use these points to find a constant, k, relating atomic mass and specific heat. Complete Table 5-7 as a guide for finding the constant.

Atomic Mass Coordinate	Specific Heat Coordinate	k	
		Atomic Mass × sp ht =	Atomic Mass / sp ht =

Table 5–7: Search for a constant relationship between atomic mass and specific heat.

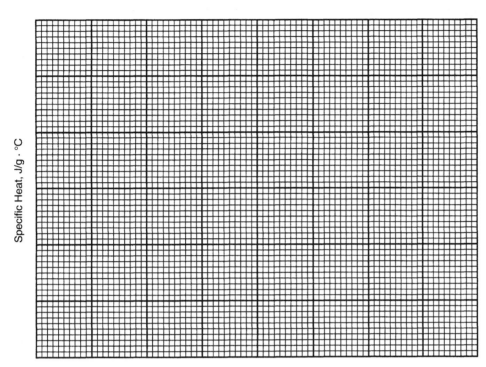

Graph 5–3: Specific heat and atomic mass of metals.

Interpretation

1. Which of the columns in Table 5-7, atomic mass × sp ht or atomic mass/specific heat produces the most constant relationship, k?

2. Which mathematical formula atomic mass × sp ht = k or atomic mass/sp ht = k is the best representation of the relationship found in step 1?

3. What is the average value of the constant, k, in the column that gave the most constant relationship?

4. Rewrite the formula derived in step 2, replacing the constant, k, with the average value found in step 3.

■ Investigation 4

Objective

To use the relationship developed to determine the approximate atomic mass of an unknown metal.

Procedure

1. Using the procedure developed in Investigation 2, determine the specific heat of the unknown metal assigned to you.

2. Record your data in Table 5-8.

Identification of Unknown Metal Used				
Mass of Metal Used		g	**Temperature of Cold Water & Metal**	°C
Mass of Hot Water Used		g		
Mass of Cold Water Used		g	**Temperature of Hot Water**	°C
Lapse Time After Mixing, s	**Temperature °C**	**Lapse Time After Mixing, s**	**Temperature °C**	
0		210		
30		240		
60		270		
90		300		
120		330		
150		360		
180		390		
Predicted Highest Temperature From Extrapolation				°C

Table 5–8: Unknown metal and water data.

3. Using the information in Table 5-8, prepare Graph 5-4.

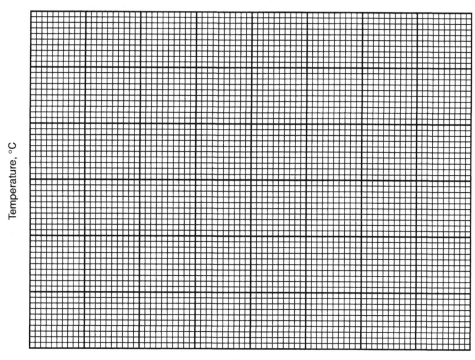

Time, s

Graph 5–4: Time-temperature plot for water and unknown metal.

Interpretation

1. Using the information recorded in Table 5-8 calculate the specific heat of the metal. Use Table 5-9 as a guide.

Identification of the Unknown Metal Used			
	Initial Temperature °C	Equilibrium Temperature °C	Temperature Change ΔT
Hot Water			
Cold Water			
Metal			
Heat Gained		=	Heat Lost
Cold Water	+	Metal =	Hot Water
sp ht$_{water}$ × ΔT_{water} × g_{water}	+	sp ht$_{metal}$ × ΔT_{metal} × g_{metal} =	sp ht$_{water}$ × ΔT_{water} × g_{water}
		=	
sp ht$_{water}$ = 4.186 J/g · °C		sp ht$_{metal}$ =	J/g · °C

Table 5–9: Calculation of the specific heat of unknown metal.

2. Using the mathematical relationship developed in Investigation 3, predict the approximate atomic mass of the unknown metal. Show all necessary calculations.

Report Sheet

Name: _____

Experiment 5

Date: _____ Section: _____

1. In Investigation 1, how did the heat lost by the hot water compare to the heat gained by the cold water? As used in this experiment, what is the purpose of extrapolating to time zero? Explain.

2. From Investigation 2, complete the following portion of Table 5-10.

Heat Gained			=	Heat Lost
Cold Water	+	Metal	=	Hot Water
sp ht$_{water} \times \Delta T_{water} \times g_{water}$	+	sp ht$_{metal} \times \Delta T_{metal} \times g_{metal}$	=	sp ht$_{water} \times \Delta T_{water} \times g_{water}$
			=	
sp ht$_{water}$ = 4.186 J/g · °C		sp ht$_{metal}$	=	J/g · °C

Table 5–10: Calculation of the specific heat of known metal.

3. In Table 5-7, which column produced the most constant relationship and what was the average value of the constant, k?

4. What was the final formula developed for relating the specific heat and the atomic mass in Investigation 3?

5. Complete Table 5-11.

Identification of the Unknown Metal Used			
	Initial Temperature °C	Equilibrium Temperature °C	Temperature Change ΔT
Hot Water			
Cold Water			
Metal			
Heat Gained		=	Heat Lost
Cold Water	+	Metal =	Hot Water
sp ht$_{water}$ × ΔT_{water} × g_{water}	+	sp ht$_{metal}$ × ΔT_{metal} × g_{metal} =	sp ht$_{water}$ × ΔT_{water} × g_{water}
		=	
sp ht$_{water}$ = 4.186 J/g · °C		sp ht$_{metal}$ =	J/g · °C

Table 5–11: Calculation of the specific heat of unknown metal.

6. What is your predicted atomic mass of the unknown metal? Show all necessary calculations.

Extra Credit

1. If the specific heat of bismuth was experimentally determined to be 0.0293 cal/g · °C, what would be its approximate atomic mass, using your experimentally determined relationship between the specific heat and atomic mass? Use the accepted mass of bismuth from the periodic table to calculate the percent error. (1 calorie/g °C = 4.186 J/ g · °C) Show your work.

What do drag racers mean when they talk about how many "g's" they pulled?

Using Spreadsheets to Analyze Objects in Motion

Background Concepts

This experiment is intended as an introduction to the study of matter in motion and its causes. Through the collection of data and its analysis, we will clarify the various terms used and define their meaning.

Speed and Velocity

When a police officer issues a speeding citation, he writes down a number to indicate how fast you were going. The units on the number will be in "miles per hour" in this country or in "kilometers per hour" elsewhere in the world. By the use of these units, we all agree that what we call *speed* is really a ratio of distance to time:

$$\text{speed} = \frac{\text{distance}}{\text{time}}$$

In this way, the police officer has quantified your speed instead of just saying, "You were going a little too fast." It will also be used to determine the amount of fine that has to be paid.

When data such as the distance vs. time is plotted on a graph, the slope of the line represents a ratio we call speed. Although we would never say our speed is negative, it is possible for the slope of the line to be negative. If we substitute the term "velocity" for speed, then we can talk about a negative ratio. For motion in only one direction, velocity is equal to the speed ratio, except that it can be negative if you're going backward. Therefore, on a graph of distance vs. time, the *slope* of the graph is the *velocity*. This is true even when the slope of the graph is changing.

Acceleration

In drag racing, it's not how fast your car can go that's important. After all, any car can have a velocity of (say) 85 miles/hour. If you want to win the race, what's important is how quickly you can build up that velocity. To make this into a number that we can measure, we again come up with a ratio:

$$\text{acceleration} = \frac{\text{change in velocity}}{\text{time}}$$

If someone says, "My top fuel dragster will go from zero to 60 miles per hour in 1 second," they are implying a ratio of 60 miles per hour (how much the velocity changes) to time (how long it takes). Someone else could say "When I drop a penny from the Empire State Building, it gains 9.8 m/s every second." That would also be a ratio of how much the velocity changes to how much time it takes. In both examples, what is being expressed is called acceleration.

Forces and Gravity

Normally, it takes an effort to make something accelerate. It's easy, though, to let something coast. In fact, if you let something coast in outer space it will continue to coast forever. This is the essence of the law of inertia, which says that any object in motion tends to stay in motion, but to accelerate (increase or decrease) its velocity, you need to apply a force. The more massive the object is, the more inertia it has, and the larger the force will have to be. We can define how much this force needs to be by writing it as:

$$F = m\,a$$

Of course, we don't always get perfect coasting because of frictional losses such as those caused by the air or ground pushing back.

Experimentally it can be shown that when you drop a penny from a tall building, it accelerates downward, increasing its velocity by 9.8 m/s each second. We call this the acceleration of gravity, g. Since we said that accelerations are caused by forces, gravity must constantly be applying a force on the penny. If we could just turn off gravity for an instant, how hard would we have to pull on the penny to just exceed its inertia and make it accelerate at 9.8 m/s each second?

Application of the formula $F = ma = m$ (9.8 m/s^2) tells how hard. So the force of gravity, call it W for "weight," must also be that strong or $W = m\,g$. This force is present all the time, pulling the penny down even if the object can't accelerate – like when it's in your pocket.

Investigation 1

Objectives

To determine the velocity of a forward-moving object, plot a graph of the object's position as a function of time, and determine the slope of the best fit line on the graph.

Procedure

1. Your instructor will provide you with a motorized object that moves forward when you set it in motion.

2. Lay 2 meter sticks parallel to each other as guides for the object so that it will travel straight as shown in Figure 6-1.

3. Lay a ruler across the meter sticks so that the edge closest to the toy lies at 20 cm.

4. Set the toy in motion, starting with its front at the 0 cm mark. Using a stopwatch, measure how long it takes to hit the edge of the ruler that is at the 20 cm point. Repeat and measure the time for this distance 3 times. Record your data in Table 6-1.

5. Repeat the above procedure for distances of 40 cm, 60 cm, 80 cm, and 100 cm. Record your data in Table 6-1.

Figure 6-1

Distance to Ruler	Time to Travel	
20 cm	Trial #1: _____ s Trial #2: _____ s Trial #3: _____ s	 Average time: _____ s
40 cm	Trial #1: _____ s Trial #2: _____ s Trial #3: _____ s	 Average time: _____ s
60 cm	Trial #1: _____ s Trial #2: _____ s Trial #3: _____ s	 Average time: _____ s
80 cm	Trial #1: _____ s Trial #2: _____ s Trial #3: _____ s	 Average time: _____ s
100 cm	Trial #1: _____ s Trial #2: _____ s Trial #3: _____ s	 Average time: _____ s

Table 6-1: Distance vs. travel time.

Interpretation

I. Graphing:

> **Note:** Although there are other spreadsheets that may be used for this part of the experiment, the directions that follow are specific to Microsoft Excel® software.

Plot the data collected in Table 6-1 on a graph of distance vs. time, as follows:

1. On your computer, start Microsoft Excel®. You should have an empty worksheet on your screen.

2. In the first column (Column A), enter your results for the average time as shown in Figure 6-2. This will correspond to the *x*-components of the data points to be plotted.

3. In the second column (Column B) enter the distances corresponding to each of these times. These values will correspond to the *y*-components of the points to be plotted.

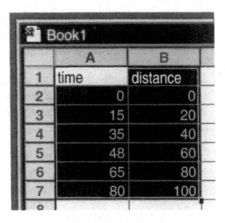

Figure 6-2

4. By clicking and dragging the mouse over the two columns, select both columns.

5. Under the "Insert" menu, select "Insert Chart."

6. As shown in Figure 6-3, choose as your chart type the "XY (Scatter Plot)." This will produce a graph whose points get their *x* and *y* coordinates from your data.

7. In the NEXT dialog box, make sure that your data series is to be interpreted in COLUMNS. The data range that's listed should correspond to the cells you selected in step 4. Normally, this is also shown by a blinking line drawn around your data.

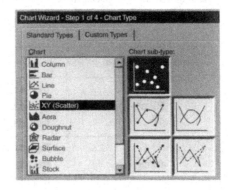

Figure 6-3

8. In the NEXT dialog box, (see Figure 6-4) you will have the option to tailor your graph. Click on the TITLES tab, if it isn't already selected. This gives us the choice to give names to the graph, the *x*-axis,

Figure 6-4

and the *y*-axis. Since we are putting distance on the *y*-axis, enter "Distance (cm)" as the title for your *y*-axis. Similarly, put "Time (seconds)" as the title of your *x*-axis. You should also give your graph a title, like "Motion of a Motorized Toy."

9. Click on the LEGEND tab of this dialog box. Since we are only plotting one set of data, we don't need a legend to help distinguish one set of points from another. Click on the check box so that the Legend is not shown.

10. Click on the GRIDLINES tab of this dialog box. Click on the different checkboxes to try out different sets of gridlines. These are the horizontal and vertical reference lines that appear on the graph to help you later read the values. Select the ones that make your graph readable, without adding unnecessary clutter.

11. In the NEXT dialog box, the computer asks where to put the chart. Choose to put it as an object in the current sheet.

12. When your graph appears, click on its corner, stretch it, and move it so that it fits beside your data.

II. Analysis of Data

1. While the graph is on your screen, select "Add Trendline…" You can find this option in the CHART menu.

2. Using the menu, add a trendline that is <u>linear</u>. Set its <u>options</u> so that it has an <u>intercept set to zero</u>, and so that the equation of the trendline appears on the chart. This will plot a straight line on the graph that best agrees with the data points. The line might not touch every data point, but it should appear as a reasonable compromise between all of the points.

   Are there any data points that are far away from the line?

3. Along with the line, there should have appeared an equation for the line. Move that equation to the top of the graph, so that it can be seen more clearly. Write the equation below:

$$y = \underline{\hspace{3cm}}$$

4. Since the computer doesn't really know what your data represents, it uses the symbols "y" and "x" in the equation to represent the quantities that you're plotting on the vertical axis, and the horizontal axis, respectively. Since you have plotted DISTANCE on the vertical axis and TIME on the horizontal axis, we know that *y* is really DISTANCE and *x* is TIME. Re-write the equation in step 3 in terms of DISTANCE and TIME below.

5. The number in the above equation is the slope of the line on the graph. Using this number, determine the average velocity of the toy. Include units in your answer.

$$v = \underline{\hspace{2cm}}$$

6. Because this number was calculated for a line that is supposed to fit all the data points, it might not agree with each data point completely. To see how big a difference this makes, let's calculate the velocity of the toy based on only two data points:

   a. How long did it take the toy to reach a distance of 20 cm?

   $$t_{20} = \underline{\hspace{2cm}} \text{ seconds}$$

   b. How long did it take the toy to reach a distance of 60 cm?

   $$t_{60} = \underline{\hspace{2cm}} \text{ seconds}$$

   c. The average velocity of the toy between these two positions should be given by:

   $$v_{20-60} = \frac{\Delta x}{\Delta t_{20-60}}$$

   $$v_{20-60} = \frac{60 \text{ cm} - 20 \text{ cm}}{t_{60} - t_{20}}$$

   $$v_{20-60} = \frac{60 \text{ cm} - 20 \text{ cm}}{\underline{\hspace{1cm}} - \underline{\hspace{1cm}}} = \underline{\hspace{2cm}} \text{ cm/s}$$

7. Suppose your measurement for t_{60} is off by 10%. Do the same calculation, but multiply your value of t_{60} by 1.1 first. How does the above velocity change?

8. Now go back to the data in your spreadsheet and alter the value of t_{60} in the appropriate cell. This should cause your line to shift slightly, and your equation to change. As a result of this, how does the velocity that you got for step 5 above change?

9. Why do you suppose it is a good idea to measure the velocity using the slope of a trendline for many data points, as opposed to using two individual data points?

10. Return the value of t_{60} to its original. Using the equation from your trendline, predict how far the toy would go in 60 seconds.

11. Using the equation from your trendline, calculate how long it would take the toy to travel 51 cm.

12. Print out a copy of your graph, along with the data beside it. Attach it to your final report sheet.

Investigation 2

Objective

To measure the position of an accelerating object over time.

Procedure

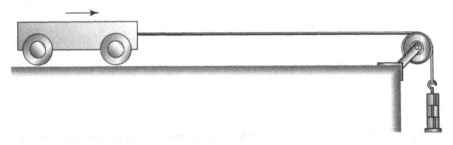

Figure 6-5

1. Set up a cart so that it is pulled by a hanging weight on a string as shown in Figure 6-5.

2. The hanger should pull the cart forward when you let it go. As a result, the cart should pick up speed. Adjust the mass of the hanger so that the cart covers a distance of 80 cm in about 1 – 2 seconds.

   mass of cart used: _____

   mass of hanger used: _____

3. Lay a meter stick beside the path of the cart to measure the distance it travels.

4. Lay a ruler across the path of the cart so that its nearest edge lies 5 cm in front of the cart.

5. Let the cart go, starting with its front in line with the 0 cm mark on the meter stick. Using a stop watch, measure how long it takes to hit the edge of the ruler that's at 5 cm. Measure this time 3 times. Record your data in Table 6-2.

6. Repeat the procedure for distances of 10 cm, 20 cm, 40 cm, 60 cm, and 80 cm. Record your data in Table 6-2.

Distance to Ruler	Time to Travel	
5 cm	Trial #1: _____ s Trial #2: _____ s Trial #3: _____ s	 Average time: _____ s
10 cm	Trial #1: _____ s Trial #2: _____ s Trial #3: _____ s	 Average time: _____ s
20 cm	Trial #1: _____ s Trial #2: _____ s Trial #3: _____ s	 Average time: _____ s
40 cm	Trial #1: _____ s Trial #2: _____ s Trial #3: _____ s	 Average time: _____ s
60 cm	Trial #1: _____ s Trial #2: _____ s Trial #3: _____ s	 Average time: _____ s
80 cm	Trial #1: _____ s Trial #2: _____ s Trial #3: _____ s	 Average time: _____ s

Table 6–2: Distance vs. time for cart.

Interpretation

The interpretation for Investigation 2 is included in the interpretation for Investigation 3.

▌ Investigation 3

Objectives

To simulate the acceleration of a cart using a spreadsheet and simultaneously plot on a graph of distance vs. time the results of this simulated acceleration along with the data points from Investigation 2.

Procedure

We will again use Excel® to calculate how an accelerating object's distance changes:

1. Open a new Excel® file on your computer. Starting with the first row and moving to the right, type the following headings: *time, acceleration, v-before, v-after, distance, and data.* (One word in each cell.)

2. In the second row in the *time* column, enter "0." Our time is set to zero for the start of the motion.

3. In the second row in the *acceleration* column, type 0.5. This is the magnitude of our acceleration (0.5 m/s^2).

4. In the second row in the column, *v-before,* enter "0." Do the same for the column, *v-after.* Our object starts off at rest, with a velocity of zero.

5. In the second row in the column under *distance,* enter "0" to designate that we're starting at zero distance. Leave the *data* column blank for now.

6. In the third row of the *time* column, enter the formula =A2+0.01. This should take the starting value of time from the cell above it, and add 0.01 seconds to it. In other words, we'll be simulating the acceleration in increments of 0.01 seconds. You can think of these as "tics" of a clock hand. Each "tic" is 1/100 of a second.

7. In the third row in the *acceleration* column, enter the formula: =B2. This should copy the value of *acceleration* from the cell above it. Since we're simulating a constant acceleration, the value of the acceleration for this split second is the same as it was during the previous "tic."

8. In the third row in the *v-before* column, enter the formula: =D2. This designates that the velocity of the object right before this "tic" is the same as what it was right after the last "tic." (It copies in the value of *v-after* from the previous tic).

9. In the third row in the *v-after* column, enter the formula: =C3+(B3*0.01). This tells the computer that the velocity of the object right after this "tic" should be faster than the velocity before the "tic." Since the acceleration (B3) is 0.5 m/s every second, the velocity should increase by 0.05 m/s during this 0.01s tic. (0.05 m/s = 0.5 m/s^2 × 0.01s).

10. In the third row, in the *distance* column, enter the formula: =E2+(0.01*average (C3:D3)). This tells the computer that during this "tic," the object's distance from the starting point increases from its previous value (E2) by an amount $\Delta d = v_{average} \times 0.01$ seconds. The average velocity during this "tic" is the average of *v-before* and *v-after.*

11. Using your mouse, select the third row of entries you just made (columns AB-CDE); click on the selection with your right mouse button, and a pop-up menu as shown in Figure 6-6 will appear for you to COPY the selection.

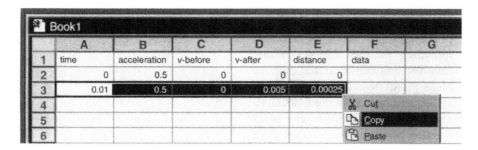

Figure 6-6

12. Now, select all the cells under those columns, down to row 200. Using the same pop-up menu, PASTE the formula entries into all those cells. Since the formulas all refer to cells above and beside them, the same calculations will be repeated all the way down, as the computer calculates the distance for each "tic."

Graphing

1. Look up your results from Investigation 2. Enter the results in the *data* column by typing in each distance value in the appropriate cell. (You might have to round your time measurements to the nearest values that fit the table). As you type in the distances, convert them to meters. So, for example, if the cart in Investigation 2 had reached a distance of 10 cm in 0.15 seconds, you should type in "0.1" in the *data* column, in the row that corresponds to a time of 0.15 seconds (row 17).

2. Starting with cell A1, select the block of cells all the way to your last entry in the data column. (It's ok to leave out the rows below this point).

3. As in Investigation 1, INSERT a chart that's an "XY (Scatter) plot." When the second dialog box pops up, click on the SERIES tab. This will give you the choice of whether to use each column as a series of data for the graph. As shown in Figure 6-7, remove the ones for *acceleration*, *v-before*, and *v-after*, since we don't want to plot those on the graph. We only want to plot the *distance* numbers from our simulation, along with the *data* from Investigation 2.

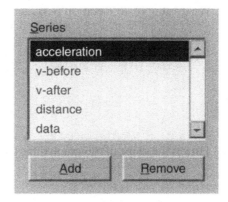

Figure 6-7

4. Proceed with laying out the graph as you did in Investigation 1. If you want, you might leave the legend displayed on the graph so you can distinguish the simulated *distance* curve from the actual *data* points.

5. Use a line to draw out the results of our simulation, rather than a jumble of points. Click on the points corresponding to the *distance* results from the simulation, so that they are highlighted. Select "Format Selected Data Series" from the FORMAT menu. When the appropriate menu window appears, select an appealing style, color and weight (thickness) of line to use, and set it so that there are no markers for this series. That will get rid of the points that are all jumbled together and only display a smooth line as shown in Figure 6-8.

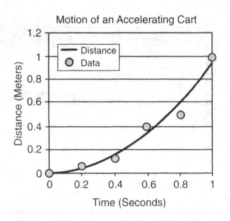

Figure 6-8

Interpretation

1. The line for the simulated distance results corresponds to an acceleration of 0.5 m/s^2, because that's what it was set for in Step 3 of the Procedure. However, this probably doesn't match the data points from Investigation 2. Go back to cell B2 and type in a different value, say 0.9. The rest of the simulated results should change automatically, as will the line on the graph.

   Is it now closer? Keep adjusting this value, while observing the changes on the graph and in the distance column. In this way, find a value for the acceleration that makes the line on the graph match your points from Investigation 2. (You can also compare the values from the *distance* column with those of the *data* column until they come close to agreeing.)

2. What value of acceleration best matches your data from Investigation 2?

   $a =$ _____ m/s^2

3. How is the shape of this graph different from that of the graph from Investigation 1?

4. How strong was the force of gravity pulling on the hanger in Investigation 2?

$$F = m_{hanger}\, g = \text{_____ newtons}$$
(m must be in kg, to get newtons)

5. How fast would this force have made the hanger accelerate if the extra inertia of the cart hadn't been there to overcome?

6. That would be the acceleration if the force in step 4 only had the inertia of the hanger itself to overcome. Instead, how much inertia was there to overcome in the cart and hanger?

$$m_{total} = m_{cart} + m_{hanger} = \text{_____}$$

7. While there is all this inertia to overcome, only m_{hanger} is involved in determining the magnitude of the accelerating force in step 4. This is only a fraction of the total mass.

a. What fraction of the total mass is it that's providing the force in step 4?

b. What fraction of g do we actually get in step 2?

c. Are these two fractions reasonably the same?

d. What factors might contribute to making (b) different from (a)?

8. Print out a copy of the graph for the report sheet.

Report Sheet

Name: _____

Experiment 6

Date: _____ **Section:** _____

1. Based on Investigation 1, what was the average velocity of the toy?

$$v_{average} = \text{_____}$$

2. For Investigation 2, report:

   a. mass of cart: _____

   b. mass of hanger: _____

   c. average acceleration: _____

   Attach your graphs for Investigations 1 and 2.

3. If the toy in Investigation 1 had moved faster, how would that have affected the line on your graph?

4. Suppose that in Investigation 2, you used a 500 g cart with a 500 g hanger, so that half of the total mass hangs over the pulley. What would you expect the acceleration to be? Explain your reasoning.

5. When a dragster accelerates at a rate of 3 g's, how much velocity does it gain every second?

Extra Credit

1. If a car accelerates at a rate of 4 m/s^2, how far will it be after 1 second, 2 seconds, and 3 seconds? Also calculate how fast it will be traveling at the end of each second.

2. A car starts with a velocity of 8 m/s, and slows down with an acceleration of –2 m/s^2. Calculate how far it will have traveled after each second, until its velocity reaches zero.

What do drag racers mean when they talk about how many "g's" they pulled?

Objects in Motion

Background Concepts

This experiment is intended as an introduction to the study of matter in motion and its causes. Through the collection of data and its analysis, we will clarify the various terms used and define their meaning.

Speed and Velocity

When a police officer issues a speeding citation, he writes down a number to indicate how fast you were going. The units on the number will be in "miles per hour" in this country or in "kilometers per hour" elsewhere in the world. By the use of these units, we all agree that what we call *speed* is really a ratio of distance to time:

$$\text{speed} = \frac{\text{distance}}{\text{time}}$$

In this way, the police officer has quantified your speed, instead of just saying, "You were going a little too fast." It will later be used to determine the amount of fine that has to be paid.

When data such as the distance vs. time is plotted on a graph, the slope of the line represents a ratio we call speed. Although we would never say our speed is negative, it is possible for the slope of the line to be negative. If we substitute the term "velocity" for speed, then we can talk about a negative ratio. For motion in only one direction, velocity is equal to the speed ratio, except that it can be negative if you're going backward. Therefore, on a graph of distance vs. time the *slope* of the graph is the *velocity*. This is true even when the slope of the graph is changing.

Acceleration

In drag racing, it's not how fast your car can go that's important. After all, any car can have a velocity of (say) 85 miles/hour. If you want to win the race, what's important is how quickly you can build up that velocity. To make this into a number that we can measure, we again come up with a ratio:

$$\text{acceleration} = \frac{\text{change in velocity}}{\text{time}}$$

If someone says, "My top fuel dragster will go from zero to 60 miles per hour in 1 second." they are implying a ratio of 60 miles per hour (how much the velocity changes) to time (how long it takes). Someone else could say "When I drop a penny from the Empire State Building, it gains 9.8 m/s every second." That would also be a ratio of how much the velocity changes to how much time it takes. In both examples, what is being expressed is called acceleration.

Forces and Gravity

Normally, it takes an effort to make something accelerate. It's easy, though, to let something coast. In fact, if you let something coast in outer space it will continue to coast forever. This is the essence of the law of inertia, which says that any object in motion tends to stay in motion, but to accelerate (increase or decrease) its velocity, you need to apply a force. The more massive the object is, the more inertia it has, and the larger the force will have to be. We can define how much this force needs to be by writing it as:

$$F = m\,a$$

Of course, we don't always get perfect coasting because of frictional losses such as from the air or ground pushing back.

Experimentally it can be shown that when you drop a penny from a tall building it accelerates downward, increasing its velocity by 9.8 m/s each second. We call this the acceleration of gravity, g. Since we said that accelerations are caused by forces, gravity must constantly be applying a force on the penny. If we could just turn off gravity for an instant, how hard would we have to pull on the penny, to just exceed its inertia and make it accelerate at 9.8 m/s each second?

Application of the formula $F = ma = m\,(9.8 \text{ m/s}^2)$ tells how hard. So, the force of gravity, call it W for "weight," must also be that strong or $W = m\,g$. This force is present all the time, pulling the penny down even if the object can't accelerate – like when it's in your pocket.

Investigation 1

Objectives

To determine the velocity of a forward-moving object, plot a graph of the object's position as a function of time, and determine the slope of the best fit line on the graph.

Procedure

1. Your instructor will provide you with a motorized object that moves forward when you set it in motion.

2. Lay two meter sticks parallel to each other as guides for the object so that it will travel straight as shown in Figure 7-1.

3. Lay a ruler across the meter sticks so that the edge closest to the toy lies at 20 cm.

4. Set the toy in motion, starting with its front at the 0 cm mark. Using a stopwatch, measure how long it takes to hit the edge of the ruler that is at the 20 cm point. Repeat and measure the time for this distance 3 times. Record your data in Table 7-1.

Figure 7-1

5. Repeat the above procedure for distances of 40 cm, 60 cm, 80 cm, and 100 cm. Record your data in Table 7-1.

Distance to Ruler	Time to Travel	
20 cm	Trial #1: _____ s Trial #2: _____ s Trial #3: _____ s	 Average time: _____ s
40 cm	Trial #1: _____ s Trial #2: _____ s Trial #3: _____ s	 Average time: _____ s
60 cm	Trial #1: _____ s Trial #2: _____ s Trial #3: _____ s	 Average time: _____ s
80 cm	Trial #1: _____ s Trial #2: _____ s Trial #3: _____ s	 Average time: _____ s
100 cm	Trial #1: _____ s Trial #2: _____ s Trial #3: _____ s	 Average time: _____ s

Table 7–1: Distance vs. travel time.

Interpretation

From the data recorded in Table 7-1, plot your position data on Graph 7-1 of distance vs. time.

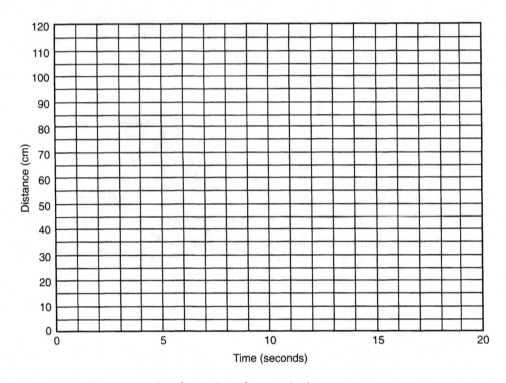

Graph 7–1: Distance vs. time for motion of a motorized toy.

1. Using a ruler, draw a straight line that starts at the origin of the graph and passes through the data points. The line should represent a compromise between all of the points, even if it doesn't actually touch all or any of the points. A good rule of thumb is that your line should have as many data points above it as below it. Extend your line all the way to the end of the graph.

2. Determine the slope of the graph using the following equation. Use its beginning and ending points where the line hits the edge of the graph:

$$\text{slope} = \frac{\text{rise}}{\text{run}} = \frac{\text{ending distance} - \text{initial distance}}{\text{ending time} - \text{initial time}}$$

$$= \frac{\underline{\hspace{2cm}} - 0 \text{ cm}}{\underline{\hspace{2cm}} - 0 \text{ seconds}} = \underline{\hspace{2cm}} \text{ cm/s}$$

3. This slope should correspond to the average velocity of the toy over the 1 meter range. Because this number was calculated for a line that is supposed to fit all the data points, it might not agree with each data point completely. To see how big a difference this makes, let's calculate the velocity of the toy based on only two data points:

   a. How long did it take the toy to reach a distance of 20 cm?

$$t_{20} = \underline{\hspace{2cm}} \text{ seconds}$$

   b. How long did it take the toy to reach a distance of 60 cm?

$$t_{60} = \underline{\hspace{2cm}} \text{ seconds}$$

   c. The average velocity of the toy between these two positions should be:

$$v_{20-60} = \frac{\Delta x}{\Delta t_{20-60}}$$

$$v_{20-60} = \frac{60 \text{ cm} - 20 \text{ cm}}{t_{60} - t_{20}}$$

$$v_{20-60} = \frac{60 \text{ cm} - 20 \text{ cm}}{\underline{\hspace{1cm}} - \underline{\hspace{1cm}}} = \underline{\hspace{2cm}} \text{ cm/s}$$

4. Suppose your measurement for t_{60} is off by 10%. Do the same calculation, but multiply your value of t_{60} by 1.1 first. How does the above velocity change?

5. Why should we expect that the result from step 2 would be less affected by your being 10% off in this single measurement of t_{60}?

6. Based on your graph, how far did the toy move in 51 seconds?

7. Using the equation distance = velocity × time, predict how far the toy would go in 60 seconds.

Investigation 2

Objective

To measure the position of an accelerating object over time.

Procedure

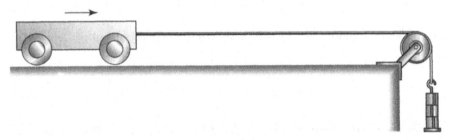

Figure 7-2

1. Set up a cart so that it is pulled by a hanging weight on a string as shown in Figure 7-2.

2. The hanger should pull the cart forward when you let it go. As a result, the cart should pick up speed. Adjust the mass of the hanger so that the cart covers a distance of 80 cm in about 1 – 2 seconds.

   mass of cart used: _____

   mass of hanger used: _____

3. Lay a meter stick beside the path of the cart to measure the distance it travels.

4. Lay a ruler across the path of the cart so that its nearest edge lies 5 cm in front of the cart.

5. Let the cart go, starting with its front in line with the 0 cm mark on the meter stick. Using a stop watch, measure how long it takes to hit the edge of the ruler that's at 5 cm. Measure this time 3 times. Record your data in Table 7-2.

6. Repeat the procedure for distances of 10 cm, 20 cm, 40 cm, 60 cm, and 80 cm. Record your data in Table 7-2.

Distance to Ruler	Time to Travel	
5 cm	Trial #1: _____ s Trial #2: _____ s Trial #3: _____ s	Average time: _____ s
10 cm	Trial #1: _____ s Trial #2: _____ s Trial #3: _____ s	Average time: _____ s
20 cm	Trial #1: _____ s Trial #2: _____ s Trial #3: _____ s	Average time: _____ s
40 cm	Trial #1: _____ s Trial #2: _____ s Trial #3: _____ s	Average time: _____ s
60 cm	Trial #1: _____ s Trial #2: _____ s Trial #3: _____ s	Average time: _____ s
80 cm	Trial #1: _____ s Trial #2: _____ s Trial #3: _____ s	Average time: _____ s

Table 7–2: Distance vs. time for cart.

Interpretation

The interpretation for Investigation 2 is included in the interpretation for Investigation 3.

■ Investigation 3

Objective

To calculate the distance and time data for an object undergoing a constant acceleration and plot a graph of distance vs. time for an object undergoing a constant acceleration.

Procedure

1. From Investigation 2, how long did it take the cart to reach a distance of 80 cm?

   $t =$ _____ s

   Using this time, determine the average acceleration:

   $$0.80 \text{ meter} = \tfrac{1}{2} \, a \, t^2$$

   $a =$ _____ m/s^2

2. We will use this value of a to compute a table of values for distances the cart reached at different times. The table updates the motion of the cart in increments or "tics" of 0.2 seconds. Since the acceleration is to remain constant, enter its value in Table 7-3 as being the same at every "tic."

3. The velocity immediately before each "tic" is the same as the velocity immediately following the previous "tic," so *v-before* for one "tic" is equal to the *v-after* from the preceding "tic." Using this, update the value of *v-before* in Table 7-3.

4. During each "tic," the velocity increases by an amount ($a \times 0.2$s). Use this to update the value of velocity immediately after each "tic" and record it in Table 7-3.

5. During each "tic," the carts distance increases by an amount ($v_{average} \times 0.2$s), where $v_{average}$ is the average of the velocities at the beginning and the end of "tic."

6. Continue down Table 7-3, updating the velocities and distances for the cart. Stop when the distance passes 0.80 meters.

Graphing

1. For each value of distance in Table 7-3, plot a small point on Graph 7-2 of distance vs. time. Draw a curve that passes *smoothly* through all these calculated points. (Don't just play connect the dots).

2. Using your results from Investigation 2, plot somewhat larger, circular points on the same graph corresponding to the distances and times of the cart's actual motion. (Be careful to convert the distances to meters).

3. Does the smooth line from step 1 fit the points from step 2 reasonably well?

Time (seconds)	a (m/s²)	v-before (m/s)	v-after (m/s)	Distance (m)
0		0	0	0
0.2		*(enter value of v-after from prev. row):*	*v-before + a × 0.2s =*	*prev. dist.* $+ \frac{v_{after} - v_{before}}{2} \times 0.2s =$
0.4		*(enter value of v-after from prev. row):*	*v-before + a × 0.2s =*	*prev. dist.* $+ \frac{v_{after} - v_{before}}{2} \times 0.2s =$
0.6		*(enter value of v-after from prev. row):*	*v-before + a × 0.2s =*	*prev. dist.* $+ \frac{v_{after} - v_{before}}{2} \times 0.2s =$
0.8		*(enter value of v-after from prev. row):*	*v-before + a × 0.2s =*	*prev. dist.* $+ \frac{v_{after} - v_{before}}{2} \times 0.2s =$
1.0		*(enter value of v-after from prev. row):*	*v-before + a × 0.2s =*	*prev. dist.* $+ \frac{v_{after} - v_{before}}{2} \times 0.2s =$
1.2		*(enter value of v-after from prev. row):*	*v-before + a × 0.2s =*	*prev. dist.* $+ \frac{v_{after} - v_{before}}{2} \times 0.2s =$
1.4		*(enter value of v-after from prev. row):*	*v-before + a × 0.2s =*	*prev. dist.* $+ \frac{v_{after} - v_{before}}{2} \times 0.2s =$
1.6		*(enter value of v-after from prev. row):*	*v-before + a × 0.2s =*	*prev. dist.* $+ \frac{v_{after} - v_{before}}{2} \times 0.2s =$
1.8		*(enter value of v-after from prev. row):*	*v-before + a × 0.2s =*	*prev. dist.* $+ \frac{v_{after} - v_{before}}{2} \times 0.2s =$
2.0		*(enter value of v-after from prev. row):*	*v-before + a × 0.2s =*	*prev. dist.* $+ \frac{v_{after} - v_{before}}{2} \times 0.2s =$

Table 7–3: Cart data per second.

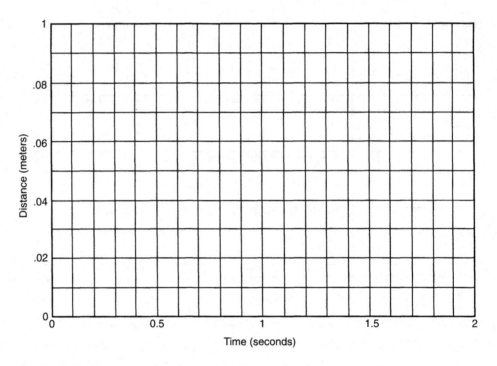

Graph 7–2: Distance vs. time for motion of an accelerating cart.

Interpretation

1. How is the shape of this graph different from the graph in Investigation 1?

2. What magnitude of acceleration does the smooth line represent?

$$a = \underline{\hspace{2cm}} \; m/s^2$$

3. How strong was the force of gravity pulling on the hanger in Investigation 2?

$$F = m_{\text{hanger}} g = \underline{\hspace{2cm}} \; \text{newtons}$$
(m must be in kg, to get newtons)

4. How fast would this force have made the hanger accelerate if the extra inertia of the cart hadn't been there to overcome?

5. That would be the acceleration if the force from step 3 only had the inertia of the hanger itself to overcome. Instead, how much inertia was there to overcome in the cart and hanger?

$$m_{total} = m_{cart} + m_{hanger} = \underline{\hspace{1.5cm}}$$

6. While there is all this inertia to overcome, only m_{hanger} is involved in determining the magnitude of the accelerating force in step 4. This is only a fraction of the total mass.

   a. What fraction of the total mass is it that's providing the force in step 4?

   b. What fraction of g do we actually get in step 2?

   c. Are these two fractions reasonably the same?

   d. What factors might contribute to making (b) different from (a)?

Report Sheet

Name: _____

Experiment 7

Date: _____ **Section:** _____

1. Based on Investigation 1, what was the average velocity of the toy?

$v_{\text{average}} =$ _____

2. For Investigation 2, report:

   a. mass of cart: _____

   b. mass of hanger: _____

   c. average acceleration: _____

   Attach the graphs you prepared for Investigations 1 and 2.

3. If the toy in Investigation 1 had moved faster, how would that have affected the line on your graph?

4. Suppose that in Investigation 2, you used a 500 g cart, with a 500 g hanger, so that half of the total mass hangs over the pulley. What would you expect the acceleration to be? Explain your reasoning.

5. When a dragster accelerates at a rate of 3 g's, how much velocity does it gain every second?

Extra Credit

1. If a car accelerates at a rate of 4 m/s^2, how far will it be after 1 second, 2 seconds, and 3 seconds? Also calculate how fast it will be traveling at the end of each second.

2. A car starts with a velocity of 8 m/s, and slows down with an acceleration of –2 m/s^2. Calculate how far it will have traveled after each second, until its velocity reaches zero.

*Can you determine which driver is telling the truth
about an auto accident?*

Momentum and Friction in a Car Crash: A Forensic Investigation

Background Concepts

Momentum

An object's momentum, p, can be calculated by multiplying its mass and its velocity: $p = m\,v$. Momentum values are negative when the object moves in the negative (opposite) direction.

Whenever an object, like a car, bumps into another object, like a truck, the car pushes off the truck and changes its momentum. But because the truck gets pushed by the car, the truck's momentum also changes. The push that the truck gets is called an impulse. The push (impulse) on the car and the push (impulse) on the truck are an action-reaction pair. So whatever impulse the truck gets, the car gets an equal impulse, but in the opposite direction. It might make the car just slow down or even bounce backward.

Figure 8-1

Conservation of Momentum

If we add up the momenta of both objects before and after they bump into each other, we should get the same total. If the car's momentum goes down, the truck's momentum goes up by the same amount, so that the total stays the same. The car's momentum might even go down so far that it becomes negative, which just means that it bounces backward. Mathematically:

$$\text{momentum before} = \text{momentum after}$$

$$m_1 v_1 + m_2 v_2 = m_1 v_1 + m_2 v_2$$

where the subscript 1 is for the car, and subscript 2 is for the truck.

Friction

An object sliding across the floor would slide forever if there were no friction. For objects sliding on a flat surface, the force of friction, f, is given by the equation:

$$f = \mu_K (mg)$$

In this equation, the force of friction is the product of the mass, m, of the object in kilograms, kg, times the gravitational force, g, (9.8 N/kg, or 9.8 m/s^2), and the coefficient of friction, μ_K. The coefficient of friction is a number that tells us how the two materials rub together. The coefficient of friction between wood and rubber is 0.4, for example.

Let's say we have a crate sliding on the floor. How far will it go before the force of friction brings it to a stop? To calculate this, we must first determine how many joules of kinetic energy the sliding crate has and equate that to how many joules of work the force of friction must do to dissipate that energy:

$$\text{initial kinetic energy} = \text{work done by friction}$$

$$\tfrac{1}{2} m \, v^2_{\text{initial}} = \text{force} \times \text{distance}$$

$$\tfrac{1}{2} m \, v^2 = f \times d_{\text{stop}}$$

Stopping distance of a sliding object, therefore, becomes:

$$d_{\text{stop}} = \tfrac{1}{2} m \, v^2 / \mu_K mg$$

This gives us a formula for the stopping distance of any object, based on its velocity and the friction conditions.

To convert miles per hour, mph, into meters per second (m/s) requires the following calculation:

$$\frac{1 \text{ mile}}{1 \text{ hr}} \times \frac{1,609 \text{ m}}{1 \text{ mile}} \times \frac{1 \text{ hr}}{3,600 \text{ s}} = 0.447 \text{ m/s}$$

Experiment Scenario:

You are hired by an insurance company to investigate a car accident that happened on the Interstate. A blue luxury sedan (BLS) ran into a red hatchback (RHB) from behind. The drivers blame each other:

Red Hatchback (RHB) (Driver A):
"I wasn't doing anything. I was in the left hand lane, going 50 mph. I looked away for one second to locate my cell phone. When I looked back, here was this blue luxury sedan in my rearview mirror coming at me real fast. He must have been doing 100 mph, for sure. I

didn't have time to do anything. He even smashed up my priceless bottle cap collection that was in the trunk."

Blue Luxury Sedan (BLS) (Driver B):
"It's not my fault. I was going 60 mph in the left-hand lane. Then this minivan in front of me swerves to the right-hand lane, real quick. That's when I see why. There was this red hatchback car stopped in front of him. I never had time to react. Man, I don't know what this guy was doing, stopped in the middle of the Interstate like that. The speed limit signs say you have to go a minimum of 40 mph on the Interstate. Now my hood ornament is all bent up."

Measurements at the site of the accident show that the vehicles locked together and left a skid mark 14 m long, from the point of impact. An expert inspected the damage to the cars and estimates that the vehicles hit at a relative velocity of about 55 ± 5 mph.

Approximate masses of the two vehicles are:

- BLS: 1,500 kg
- RHB: 1,500 kg

You must analyze the evidence in terms of momentum and friction to decide which of these stories is more likely.

Investigation 1

Objective

To become familiar with the concept of momentum and momentum transfer in elastic collisions.

Case #1: Carts of Equal Mass

Procedure

1. Adjust the track so that it is level as shown in Figure 8-2.

2. Place two carts on the track. Arrange them so that they bounce smoothly off each other.

3. Weigh the two carts and add enough mass to the lighter cart so their masses are equal. Use masking tape and slotted weights as necessary.

   Mass of each cart: _____ grams

4. Place one cart midway down the track and roll the other toward it so that it collides with the stationary cart.

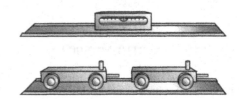

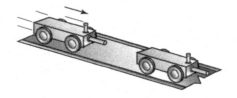

Figure 8-2

Interpretation

1. What happens to the moving cart after it hits the stationary cart?

2. How fast, in comparison to the incoming cart, does the second cart move after the collision?

3. In what way can we say that the momentum after the collision is the same as the momentum before the collision?

Case #2: Heavy Cart Hits Lighter Cart

Procedure

1. Attach extra mass to the cart being pushed so that it has at least twice the mass as the stationary cart, as shown in Figure 8-3.

2. Push the heavier cart into the less massive stationary cart.

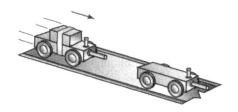

Figure 8-3

Interpretation

1. What does the heavy cart do after the collision that is different from what happened in Case #1?

2. Before the collision, how much momentum does the light cart have?

3. During the collision, the light cart gains momentum. Where do you think that momentum came from?

Case #3: Light Cart Hits Heavier Cart

Procedure

1. This time, as shown in Figure 8-4, make the light cart hit the more massive stationary cart.

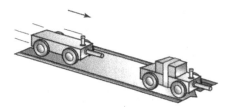

Figure 8-4

Interpretation

1. What does the light cart do after the collision?

2. How much momentum did the massive cart have before the collision?

3. Where did the momentum gained by the more massive cart come from?

4. Why can we say that the lighter cart did more than just slow down or that it gave more than it had?

Investigation 2

Objective

To gain familiarity with carts pushing off each other and how conservation of momentum applies to completely inelastic collisions.

Procedure

1. Adjust the track so that it is level.

2. Remove the extra mass so the carts again have equal mass.

3. Cock the plunger on one cart and place the carts together on the track so that when you tap the plunger release button, the plunger from one cart has to push against the other cart. See Figure 8-5.

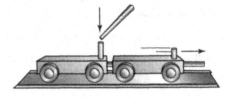

Figure 8-5

4. Make sure that there is some kind of bumper at each end of the track to stop the carts.

5. (Case #1) By trial and error, determine where you must place the carts so that after you tap the plunger, they both strike the ends of the track simultaneously. To prevent giving either cart a push, it is best to tap the button with a ruler.

6. After you have found the proper starting point, measure how far each cart travels to reach the end of the track. Record your measurements in Table 8-1.

7. Using a stopwatch, measure how long it takes the carts to reach the ends of the track. Repeat the measurement 3 times and calculate the average time. Record your data in the Table 8-1.

8. (Case #2) Add mass to one of the carts, to make it twice as massive as the other, and repeat the procedure. Record your data in the Table 8-1.

9. (Case #3) Add enough more mass to the cart to make it three times as massive as the other and repeat the procedure. Record your data in Table 8-1.

Mass of Cart #1, m_1	Mass of Cart #2, m_2	Distance Cart #1, d_1	Distance Cart #2, d_2	Time to reach the end of the track, t
$m_1 \approx m_2$: $m_1 =$ ____ g	$m_2 =$ ____ g	$d_1 =$ ____ cm	$d_2 =$ ____ cm	Trial #1: $t =$ _____ s Trial #2: $t =$ _____ s Trial #3: $t =$ _____ s Average: $t =$ _____ s
$m_1 \approx 2\, m_2$: $m_1 =$ ____ g	$m_2 =$ ____ g	$d_1 =$ ____ cm	$d_2 =$ ____ cm	Trial #1: $t =$ _____ s Trial #2: $t =$ _____ s Trial #3: $t =$ _____ s Average: $t =$ _____ s
$m_1 \approx 3\, m_2$: $m_1 =$ ____ g	$m_2 =$ ____ g	$d_1 =$ ____ cm	$d_2 =$ ____ cm	Trial #1: $t =$ _____ s Trial #2: $t =$ _____ s Trial #3: $t =$ _____ s Average: $t =$ _____ s

Table 8–1: Investigation 2 data.

Interpretation

1. When the two carts were together, we can say that they formed one large object, with a mass of ($m_1 + m_2$). What was the momentum of that (combined) object before each plunger was released?

$$p_{\text{before}} = (m_1 + m_2) \times v_{\text{before}} = \underline{\qquad}$$

(v_{before} is the velocity of the combined object before the collision).

2. For all three cases, calculate the momentum of each of the two carts to show that they are equal and opposite. Record your results in Table 8-2.

$$p_{\text{cart}} = m \times v = m_{\text{cart}}\,(d/t)$$

	Cart #1		Cart #2	
	$v_1 = d/t$	p_1	$v_2 = d/t$	p_2
Case #1: $m_1 \approx m_2$	cm/s	g cm/s	cm/s	g cm/s
Case #2: $m_1 \approx 2m_2$	cm/s	g cm/s	cm/s	g cm/s
Case #3: $m_1 \approx 3m_2$	cm/s	g cm/s	cm/s	g cm/s
	These numbers should be positive, because cart #1 goes in the forward direction		These numbers should be negative because cart #2 goes in the reverse direction.	

Table 8–2: Cart data.

3. How can we say that the total momentum is still zero after the plunger is released?

4. In the second case, the first cart should have a mass that's about twice as much as that of cart #2. Is the ratio of their distances also close to "2"? Explain.

5. What does that say about how cart #1's velocity compares to that of cart #2? Explain.

6. Now let's imagine that the two carts from case 2 had been travelling together at 30 cm/s, in the positive direction before we released the plunger. How do you suppose that would change the cart's velocities for case 2?

$$\text{new } v_1 = \underline{\hspace{1.5cm}} \text{ cm/s} + 30 \text{ cm/s} = \underline{\hspace{1.5cm}} \text{ cm/s}$$
$$(v_1 \text{ from table})$$

Cart now moves 30 cm/s faster in the positive direction;

$$\text{new } v_2 = \underline{\hspace{1.5cm}} \text{ cm/s} + 30 \text{ cm/s} = \underline{\hspace{1.5cm}} \text{ cm/s}$$
$$(v_2 \text{ from table})$$

Cart now moves 30 cm/s slower in the negative direction;

7. Now suppose that we videotaped the carts from question 6 pushing off each other in this way. If we played the tape backward, it would look like two carts coming from opposite directions, pushing against each other, sticking together and moving off. What would be the final velocity of the combined "wreck," in this backward videotape?

8. Let's see if your answer works, so that the total momentum of both carts together would stay the same before and after the collision. Calculate the following, using the data from question 6 for the reverse videotape.

Total momentum before the collision:

$$m_1 \quad \times \quad (\text{new } v_1) + \quad m_2 \quad \times \quad (\text{new } v_2)$$

$$\underline{\hspace{1.5cm}} \times (\underline{\hspace{1.5cm}}) + \underline{\hspace{1.5cm}} \times (\underline{\hspace{1.5cm}}) =$$

$$\underline{\hspace{3cm}} + \underline{\hspace{3cm}} = \underline{\hspace{1.5cm}} \text{ g cm/s}$$

Total momentum after the collision:

$$(m_1 + m_2) \times (30 \text{ cm/s}) = \underline{\hspace{1.5cm}} \text{ g cm/s}$$

These two numbers should be roughly the same. Are they?

9. Let's look at the accident claims in the same way:

Claim A:

    Blue Luxury Sedan: $m_1 = 1,500$ kg

                                            $v_1 = + 100$ mph = _____ m/s
                                                                  (convert)

    Red Hatchback: $m_2 = 1,500$ kg

                                            $v_2 = + 50$ mph = _____ m/s
                                                                 (convert)

Total momentum before the collision:

$m_1 v_1 + m_2 v_2$

                          _____ + _____ = _____ kg m/s

The velocity after the collision:

$$v_{\text{after} - A} = \frac{\text{total momentum}}{1,500 \text{ kg} + 1,500 \text{ kg}} = \underline{\hspace{2cm}} \text{ m/s}$$

Claim B:

    Blue Luxury Sedan: $m_1 = 1,500$ kg

                                            $v_1 = + 60$ mph = _____ m/s
                                                                  (convert)

    Red Hatchback: $m_2 = 1,500$ kg

                                            $v_2 = + 0$ mph = _____ m/s
                                                                 (convert)

Total momentum before the collision:

$m_1 v_1 + m_2 v_2$

                          _____ + _____ = _____ kg m/s

The velocity after the collision:

$$v_{\text{after} - B} = \frac{\text{total momentum}}{1,500 \text{ kg} + 1,500 \text{ kg}} = \underline{\hspace{2cm}} \text{ m/s}$$

To determine which driver is right, we need to see how well these figures match the skid marks.

▋ Investigation 3

Objectives

To determine the coefficient of friction for rubber on concrete and use that value to calculate the stopping distance.

Procedure

1. Obtain a large black rubber stopper, about 2.5 inches in diameter (#13), and a 100 mL beaker or cup. The exact sizes are not important, but the stopper should fit snugly inside the cup.

2. Place as many masses inside the cup as you can, but make sure the stopper can still fit on the cup.

3. Build a pulley system as shown in Figure 8-6, to drag the cup upside-down on its stopper.

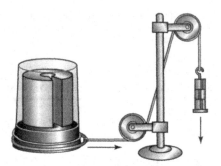

Figure 8-6

4. Move the apparatus to a clean, smooth concrete floor or other flat level surface.

5. Clear the floor of any grit or dust, and set up the apparatus with the rubber stopper dragging on the surface.

6. Add slotted masses to the weight hanger gradually. Carefully determine how much mass you need on the hanger so that when you push the cup toward the pulley, it keeps sliding for a pretty good distance on the concrete.

> **NOTE:** In order to keep the cup from skipping or stuttering, place additional mass on top of the cup. Normally, police investigators will perform a similar measurement using a drag tire pulled over the same surface traveled by the accident vehicles.

7. Using a balance, carefully measure the total mass of the cup and stopper, along with whatever mass is on top of it. Also measure the total mass it took to drag the cup, including the mass of the hanger. Record your data here:

    a. Total mass being dragged: m_{drag} = _____ grams

    b. Total mass required to pull the cup: m_{hang} = _____ grams

Interpretation

1. How much force, in newtons, was required to drag the cup at a constant velocity? ($g = 9.8$ N/kg)

$$F = m_{\text{hang}} g = \underline{\hspace{1.5cm}} \text{ newtons}$$

This is equal to the force of kinetic friction.

2. What is the coefficient of kinetic friction for rubber on concrete?

$$\mu_K = \frac{\text{force of friction}}{m_{\text{drag}}\, g} = \underline{\hspace{1.5cm}}$$

3. Right after the collision, friction would start to slow the wreck down. Using this value of μ_k, calculate the stopping distance for the wreck;

   a. if it had the velocity $v = v_{\text{after} - A}$, as it would if the first claimant was right: (This is the first velocity you calculated in step 9 of Investigation 2.)

   $$d_A = \underline{\hspace{1.5cm}} \text{ m}$$

   b. if it had the velocity $v = v_{\text{after} - B}$, as it would if the driver of the blue luxury sedan were right: (This is the second velocity you calculated in step 9 of Investigation 2.)

   $$d_B = \underline{\hspace{1.5cm}} \text{ m}$$

4. Since the police measured the skid marks to be 14 meters long, who would you say is more likely to be telling the truth?

5. How confident are you in your result? Would you feel comfortable testifying about it in court?

Report Sheet

Name: _____

Experiment 8

Date: _____ **Section:** _____

1. Based on your investigations, report:

    a. The coefficient of kinetic friction: $\mu_K = $ _____

    b. If driver "A" (RHB) were correct,

    — velocity of the wreck would be: $v_{after-A} = $ _____ m/s

    — the calculated stopping distance of the wreck: _____ m

    c. If driver "B" (BLS) were correct,

    — velocity of the wreck would be: $v_{after-B} = $ _____ m/s

    — the calculated stopping distance of the wreck: _____ m

2. Based on these findings, which of the two driver's stories is more likely to be true?

3. If the roads had been wet and slick, how would that have changed μ_K?

4. When the BLS crashed into the RHB, the RHB was accelerated (at least in the short term). How can we say that the total momentum stayed the same?

5. How much momentum did each car have immediately following the collision if driver B's statement is correct?

6. Is it true that $30 \text{ m/s} \cong 60 \text{ mph}$?

Extra Credit

1. The kinetic energy of an object in joules, is given by $K = \frac{1}{2} m v^2$. Calculate how many joules of kinetic energy each vehicle would have before the collision, if driver A's statement were correct. Does this equal the number of joules the wreck has immediately after the collision?

2. Suppose your measurement of the coefficient of friction had been off by 0.1 (either higher or lower). Recalculate your findings both ways. Does it make a significant difference in your conclusion as to who's responsible for the collision.

How can your radio pick out just one radio station at a time?

Waves and Oscillations

Background Concepts

Oscillations

For something to oscillate back and forth or up and down, it needs to have a force pulling it back toward a middle position. An example would be a mass hanging on a rubber band. If you pull the mass down, the rubber band pulls it up, but because the mass has inertia, every time the force of the rubber band pulls it up, the inertia of the mass makes it overshoot the middle position. It therefore continues to bounce up and down. The time it takes for each bounce is given by the equation

$$T = 2\pi\sqrt{\frac{m}{k}}$$

which is called the period of the oscillation. In this equation, m is the mass that's bouncing (in kg) and, k, is a constant related to the stiffness of the rubber band. We usually call this the spring constant and measure it in newtons/meter. A stiff rubber band or spring has a larger k value than a loose one.

If a mass is bouncing very fast, we may be interested in the number of oscillations per second, or the frequency, f. The frequency of a mass bouncing on a spring or rubber band is given by the equation:

$$f = \frac{1}{T} = \frac{1}{2\pi}\sqrt{\frac{k}{m}}$$

The unit for frequency is the Hertz. So, if the mass bounces 20 times per second, we say that $f = 20$ Hz. This also means that the period of each bounce is 1/20 of a second, which is $1/f$.

The amplitude of the bounce – that is, how far away from the middle position the object moves – may also be measured. In many cases, the amplitude doesn't affect the frequency or the period.

Physical Science: What the Technology Professional Needs to Know

Waves

If a long spring is stretched horizontally between two points, each point on the spring can be made to oscillate up and down, just like the mass on the rubber band. However, when you make one point on the spring move up and down, it will try to make the points beside it move up and down, too. In this way, the initial oscillation you provide will move down the spring. If you continue to move your end up and down, you will create a wave in the spring, as shown in Figure 9-1.

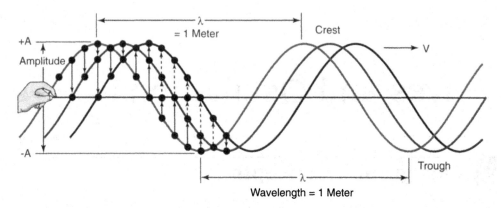

Figure 9-1

Because each point along the spring is a little slow in responding to your up and down motion, a point 0.5 m from you will still be going up as you move your end down. Even further down the spring, say 1 m, there will be a point that is going down at the same time, because it's only now doing what you did a little while ago.

If you measure the distance between your end and the point that's in sync, but one bounce behind, that distance is called the wavelength, λ. For example, if it takes 1/20 of a second (T) to get exactly one bounce behind, then we can say that the oscillation moved down the spring and covered 1 m in 1/20 of a second. That gives us the following equations for calculating the velocity of the wave:

$$v_{\text{wave}} = \frac{\lambda}{T} \quad \text{or} \quad v_{\text{wave}} = \lambda f$$

Standing Waves and Waves Reflections

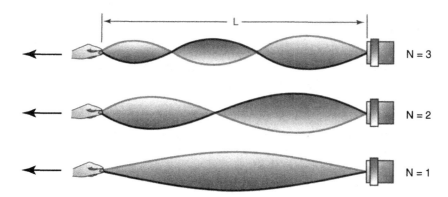

Figure 9–2: Three possible standing waves on a string of fixed length, *L*. The number of standing waves can be increased by increasing the tension on the string.

As shown in Figure 9-2, when a wave on a spring hits the end, it can bounce back or reflect the wave in one of two ways. It can either reflect upside down or right side up, depending on what's at the end of the spring. If the wave moving toward the end is trying to make the spring go up, but the wave reflecting is trying to make it go down, the two will cancel each other.

If the wave moving toward the end has just the right wavelength, the reflecting wave will be in sync with it. When the wave moving toward the end is trying to make a point in the spring go up at the same time as the reflecting wave, it creates standing waves. Standing waves look like they're neither coming nor going when, in fact, they're doing both. They are a wave and its reflection, traveling back and forth on top of each other.

Investigation 1

Objective

To determine the frequency of oscillation of a hanging mass.

Procedure

1. Hang a 300 gram mass from a rubber band, as shown in Figure 9-3.

   Exact mass hung:

   _____ grams

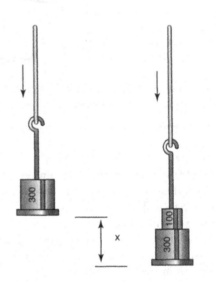

Figure 9-3

2. Measure how much the height of the hanging mass changes:

   a. when you add 100 grams to the hanging 300 g mass:

      $x = $ _____ cm or

      $x = $ _____ meters

   b. when you remove 100 grams from the hanging 300 g mass:

      $x = $ _____ cm or

      $x = $ _____ meters

3. Tape the 300 gram mass with masking tape, so that it is secure. Pull it down a few cm and let it go. Using a stop watch, measure how long it takes to complete 10 bounces. Record your results in Table 9-1. Repeat this two more times and record your results, then take an average:

$T_{10 \text{ bounces}}$	Seconds
Trial 1	
Trial 2	
Trial 3	
Average	

Table 9–1: Bounce data.

> **NOTE:** Are you really counting 10 bounces? If you start the stopwatch as soon as you let go of the mass, then 1 bounce should be counted the next time the mass is in the down position.

Interpretation

1. By how many newtons did you increase the pull on the rubber band in step 2a?

$$F = mg = (0.100 \text{ kg}) (9.8 \text{ newtons} / \text{kg}) = \underline{\hspace{1cm}} \text{ newtons}$$

2. Using your value from 2(a) in the Procedure, determine how stiff the rubber band is in terms of newtons per meter:

$$k = \frac{\underline{\hspace{1cm}} \text{ newtons}}{\underline{\hspace{1cm}} \text{ meters}} = \underline{\hspace{1cm}} \text{ N/m}$$

3. By how many newtons did you decrease the pull on the rubber band in step 2b?

$$F = mg = (0.100 \text{ kg}) (9.8 \text{ newtons} / \text{kg}) = \underline{\hspace{1cm}} \text{ newtons}$$

4. Using your value from 2b in the Procedure, determine how stiff the rubber band is in terms of newtons per meter:

$$k = \frac{\underline{\hspace{1cm}} \text{ newtons}}{\underline{\hspace{1cm}} \text{ meters}} = \underline{\hspace{1cm}} \text{ N/m}$$

5. Calculate the average value of the constant, k, from 1 and 3 in the Interpretation:

$$k = \underline{\hspace{1cm}} \text{ N/m}$$

6. Using the formula for the period of an oscillating mass, calculate what the period of oscillation should be. Use the above value of, k, in the equation, along with the exact mass you bounced on the rubber band:

$$T = \underline{\hspace{1cm}} \text{ seconds}$$

7. Based on your average T for 10 bounces, determine the time for 1 bounce:

$$T = \underline{\hspace{1cm}} \text{ seconds}$$

8. Calculate the percent deviation between your result in 7 and the theoretical result from 6:

$$\% \text{ deviation} = \left| \frac{\text{your result} - \text{theoretical result}}{\text{theoretical result}} \right| \times 100\% = \underline{\hspace{1cm}} \%$$

9. Determine the frequency of oscillation of your mass, based on your result from 7:

$$f = 1/T = \underline{\hspace{1cm}} \text{ Hz}$$

Investigation 2

Objective

To observe the characteristics of a traveling wave.

Procedure

One Oscillator

1. Obtain a 5.5 × 8.5 inch piece of acetate (half a transparency). As shown in Figure 9-4, use a marker to draw a dot in the middle of it.

2. Roll the acetate into an 8.5-inch long cylinder, roughly the same diameter as a quarter. (Hint: to prevent smudging, it helps to have the side with the dot on it facing the inside). Tape the cylinder with transparent tape.

3. Set up a metronome so that it ticks every 2 seconds.

   Alternate version: connect the two wires of a loudspeaker to the two output wires of a signal generator. Set the generator to produce a 3-volt square wave with a frequency of 0.25 Hz.

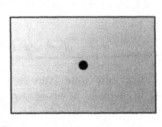

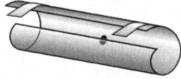

Figure 9–4: Transparency with one dot.

Now we'll use the image on the acetate as a physical representation of a traveling wave.

4. Using the metronome as a guide, roll the cylinder between your fingers as smoothly as you can, so that it completes one rotation at every click. How does the motion of the dot compare with the bouncing motion of the mass in Investigation 1?

5. Time the length of time required for the dot to complete one oscillation:

$$T = \text{_____ seconds}$$

6. Calculate the frequency. Is this equal to 0.5 Hz?

$$f = 1/T = \text{_____ Hz}$$

7. Carefully measure the diameter of the cylinder.

$$D = \text{_____ cm}$$

8. How far from its middle does the dot travel? Is this equal to $D/2$?

$$\text{amplitude} = \text{_____ cm}$$

Two Oscillators

9. Unroll the cylinder and draw a second dot on the acetate, so that the second dot is between the first dot and the corner of the acetate. (See Figure 9-5)

10. Roll the cylinder back up and repeat the oscillatory motion. Describe how the motion of the two dots is different:

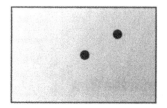

Figure 9–5: Transparency with two dots.

11. Does the second dot have

   a. The same period?   Y / N

   b. The same frequency?   Y / N

   c. The same amplitude?   Y / N

   We say that the second dot is "out of phase" with the first one if it reaches its peak at a different time.

Wave

12. Unroll the cylinder and draw more dots, as shown in Figure 9-6, all in a straight line passing through the two that you have already. Also draw two vertical lines, each 1 cm from the edge as shown in Figure 9-6.

13. Roll the cylinder back up, tape it, and continue rolling it in your fingers in sync with the metronome.

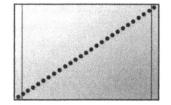

Figure 9–6: Transparency with a line of dots.

14. To determine the speed of the wave, measure:

   a. How long it takes one of the "humps" to go from one of the lines to the other:

   $$\text{travel time} = \underline{\hspace{1.5cm}} \text{ seconds}$$

   b. How far apart the two lines are:

   $$\text{travel distance} = \underline{\hspace{1.5cm}} \text{ cm}$$

   c. Calculate how fast it travels:

   $$\text{wave velocity} = \frac{\text{travel distance}}{\text{travel time}} = \underline{\hspace{1.5cm}} \text{ cm/s}$$

15. Measure the wavelength, i.e. the distance between two of the traveling "humps:"

   $$\lambda = \underline{\hspace{1.5cm}} \text{ cm}$$

16. Since every spot in the wave should have the same frequency (f) and period (T) that you measured in step 5, calculate what the velocity of the wave should be:

$$\text{wave velocity}_{\text{theory}} = \frac{\lambda}{T} = \lambda f = \underline{\hspace{2cm}} \text{ cm/s}$$

Interpretation

1. Are the answers in 15 and 16 the same? Suggest what might be the sources of error that could cause a deviation.

2. How can we say that each dot is just a little bit out of phase with the dot before it?

Investigation 3

Objective

To determine the resonant frequency of the air in a cylinder and understand how it is related to the wavelength.

Procedure

1. Connect the two wires of a loudspeaker to the two output leads of a signal generator.

2. Set the signal generator to produce a sine wave of 200 Hz, at about 3 volts. Adjust the voltage so that you can hear (softly) the sound the loudspeaker makes. Too high a voltage will damage the loudspeaker, so you might want to get your instructor's advice on how high you can safely go.

3. Take a 100 mL graduated cylinder and hold the loudspeaker over it as shown in Figure 9-7. Slowly increase the frequency by about 10 Hz at a time until you hear the sound get much louder. At what frequency does this happen? Get as precise as you can. (Note: if you get past 500 Hz, then you probably missed it).

   frequency$_1$ = _____ Hz

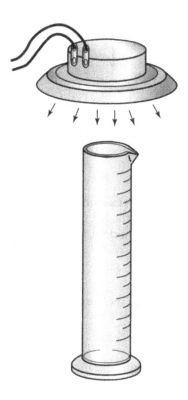

Figure 9–7: Cylinder and loud-speaker setup.

4. How long is the empty part of the cylinder?

$$L_1 = \text{_____ cm}$$

5. Now put an inch or two (10–20 mL) of water in the cylinder and try it again.

$$\text{frequency}_2 = \text{_____ Hz}$$

$$L_2 = \text{_____ cm}$$

Interpretation

1. The sound that the loudspeaker sends out is a wave. When we get it just right, it will form a standing wave inside the cylinder, bouncing back and forth from top to bottom and back. For this to work, the wave has to be at a node when it bounces off the bottom, and it has to be at an antinode when it bounces back at the top.

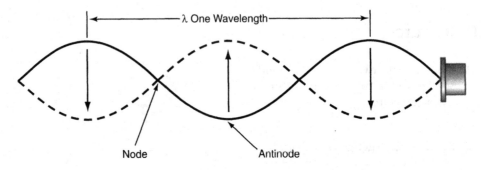

Figure 9–8

a. Let's examine the wave in Figure 9-8. The simplest way to make it work is to put the left edge (node) at the bottom of the cylinder, and put the very next antinode at the top. Measure this distance on the page:

$$d_{page} = \underline{\hspace{1.5cm}} \text{ mm}$$

b. Also measure the wavelength as it appears in this picture:

$$\lambda_{page} = \underline{\hspace{1.5cm}} \text{ cm}$$

c. If this distance, d, corresponds to the length of the tube (in step 4 above), then what wavelength does λ_{page} really correspond to?

$$\frac{d_{page}}{L} = \frac{\lambda_{page}}{\lambda}, \text{ so } \lambda = \frac{L \times \lambda_{page}}{d_{page}} = \underline{\hspace{1.5cm}} \text{ cm}$$

2. This is how long the wavelength ought to be, to make the air in the cylinder resonate. Is it the same as $4 \times L$?

3. Convert this wavelength into meters:

$$\underline{\hspace{2cm}} \text{ m}$$

4. The speed of a sound wave in the air is usually about 343 m/s. Using this, figure out what the frequency of this wave should be:

$$f = \frac{v}{\lambda} = \frac{343 \text{ m/s}}{\underline{\hspace{1cm}} \text{ m}} = \underline{\hspace{1cm}} \text{ Hz}$$

5. How does this compare to the frequency you found in step 3 of the Procedure?

6. When we put the water in the cylinder, we shortened the length of the air column in which the sound wave could exist and bounce around. Repeat your calculations using the shorter length:

$$\frac{d_{page}}{L_2} = \frac{\lambda_{page}}{\lambda_2}, \text{ so } \lambda_2 = \frac{L_2 \times \lambda_{page}}{d_{page}} = \underline{\hspace{1cm}} \text{ cm}$$

or _____ meters

$$f_2 = \frac{v}{\lambda_2} = \frac{343 \text{ m/s}}{\underline{\hspace{1cm}} \text{ m}} = \underline{\hspace{1cm}} \text{ Hz}$$

7. How does this compare with the second frequency you measured?

Report Sheet

Name: _____

Experiment 9

Date: _____ Section: _____

1. Using the data you collected, fill in the following Table:

Investigation 1: Oscillator		
$k =$ _____ N/m		
Theoretical Period for 1 bounce (T_{theory}): _____ seconds	Measured Period for 1 bounce (T_1): _____ seconds	Percent deviation: _____ %
Investigation 2: Waves		
Wavelength $\lambda =$ _____ cm		
Theoretical Wave Velocity (wave velocity$_{theory}$): _____ cm/s	Measured Wave Velocity (wave velocity): _____ cm/s	Percent deviation: _____ %
Investigation 3: Standing Wave		
Length $L_1 =$ _____ cm Wavelength $\lambda_1 =$ _____ cm		
Theoretical Frequency (f_{theory}): _____ Hz	Measured Frequency (f_1): _____ Hz	Percent deviation: _____ %
With water in it:		
Length $L_2 =$ _____ cm Wavelength $\lambda_2 =$ _____ cm		
Theoretical Frequency (f_{theory}): _____ Hz	Measured Frequency (f_2): _____ Hz	Percent deviation: _____ %

Note: Percent Deviation $= \left| \dfrac{\text{Measured Result} - \text{Theoretical Result}}{\text{Theoretical Result}} \right| \times 100\%$

2. If you softly played a mixture of different frequencies into the cylinder, what would you expect to hear?

3. A radio has a circuit that tends to oscillate back and forth electronically. Lets say that by adjusting one of the circuit elements, you make it tend to oscillate at 100 000 000 Hz. How would it respond if you sent it two radio waves at the same time:

   a. one at 100 500 000 Hz?

   b. one at 88 000 000 Hz?

4. How can a radio pick out just one station at a time?

5. If the rubber band you used in Investigation 1 had been stiffer, how would it have affected how fast the mass oscillates?

6. In Investigation 2, how can we say that the oscillators don't go in the same direction that the wave is moving?

Extra Credit

1. In question 2 above, suppose you were sending just one radio wave, but you could change its frequency from 100 000 000 Hz to 88 000 000 Hz back and forth. Describe how you could use this to send Morse code.

2. When an earthquake happens, it sends waves through the Earth, all the way across the planet. When the wave goes through the planet, its speed might change in different places, depending on what it's going through. Explain why that doesn't change its frequency, although it would change the wavelength.

How can you lift a 1-ton car by yourself?

Simple Machines

Background Concepts

Forces

A force is a push or a pull. For example, when you lift a TV you exert an upward force on it. However, while you are doing this there is also a downward force pushing the TV back down. This is the force of gravity. The force of gravity acting on an object depends on its mass. If the object has a lot of mass, the force of gravity makes that object heavy. If you want to lift the TV, the upward force needs to be stronger than the force of gravity, but not by much. If the upward force is too strong, the TV gains speed upward. Lifting a TV involves giving it an upward force that is just a little bit greater than the force of gravity, so we can control how fast it goes up. We do this without even thinking.

The force of gravity pulling down on an object is given by the formula:

$$F_G \text{ (in newtons)} = m\,g$$

where, m, is the mass of the object in kilograms and, g, is the strength of the gravitational field, 9.81 N/kg (this is also the acceleration in m/s^2 that the object would undergo if you dropped it). Usually we refer to this force as weight: $W = m\,g$.

Mechanical Advantage

Often, we use a machine to change the amount of force we're applying. That's what a machine is: a device that is used to change the strength of the force you apply. For example, a mechanic uses a system of chains and pulleys to lift an engine out of a car. He's not strong enough to lift the 300 kg engine directly. That would take (300 kg) × (9.81 N/kg) = 2,943 newtons! (The weight of about 5 men). Because of his pulley system, though, he can lift the engine with an exertion of only about 100 newtons of pull on the chain. The pulley system is a machine that takes his 100 newtons and turns it into the larger force needed to lift the engine.

A.M.A.: Definition

We say that the machine provides an actual mechanical advantage, A.M.A., which is how much the force is multiplied. In the mechanic's case, the mechanical advantage is

$$\frac{\text{force exerted by machine}}{\text{force exerted by you}} = \frac{2943 \text{ N}}{100 \text{ N}} = 29.4 = \text{A.M.A.}$$

since his 100 newtons was multiplied 29.4 times by the pulley system.

It is likely that this number would be even larger if there were no friction in the pulley system, and if the chains were not themselves being pulled down by gravity. Because of this, a distinction is made between the ideal mechanical advantage (I.M.A.), and the actual mechanical advantage (A.M.A.). The I.M.A. is the mechanical advantage that the machine would afford if there were no friction, and if the machine parts themselves had no mass (and hence no weight). It is a number calculated from the drawing board, describing how the machine was designed to act in those perfect conditions. The A.M.A., on the other hand, is what you actually get when you use the machine in the real world, friction and all. Typically, the A.M.A. is lower than the I.M.A.

Work and Energy

Another quantity scientists talk about is work. The amount of work done on the engine is defined as the force acting on it multiplied by the distance over which that force acts. So, if the engine is lifted 0.30 meters (about 1 foot), the work done on it would be:

$$\text{work} = F_{\text{engine}} \times d_{\text{engine}}$$
$$= 2943 \text{ N} \times 0.30 \text{ m}$$
$$= 880 \text{ joules}$$

Work involves the transfer of energy (joules). In this case, chemical energy from within the mechanic's body is turned into the gravitational potential energy of the suspended engine. Ideally, the work done by the mechanic in pulling the chain (with 100 N of force) should be equal to the work done on the engine (it 2,943 N of force). Then, as the law of conservation of energy states, energy is neither created nor destroyed: the mechanic loses as many joules as the engine gains. How can these two measures of work be equal? Although the mechanic pulls the chain with 29.4 times less force, he will have to pull it 29.4 times further:

$$F_{\text{mech}} \times d_{\text{mech}} = F_{\text{engine}} \times d_{\text{engine}}$$
$$d_{\text{mech}} = \frac{F_{\text{engine}}}{F_{\text{mech}}} \times d_{\text{engine}}$$
$$d_{\text{mech}} = \frac{2943 \text{ N}}{100 \text{ N}} \times 0.30 \text{ m}$$
$$= 8.8 \text{ meters}$$

He has to pull the chain 8.8 meters. One way to say this is that as far as the machine is concerned:

$$\text{work in} = \text{work out.}$$

I.M.A.: Practical Definition

The ratio of these two distances gives us a practical way of measuring the I.M.A.

$$\text{I.M.A.} = \frac{\text{distance you pull}}{\text{distance the "load" goes up}} \quad \text{(practical definition)}$$

This measure of the mechanical advantage is nearly ideal, because the distances are the same with or without friction. As long as the chains don't stretch, this should give the same number as the ideal value

Efficiency

Although the work done by the mechanic should be the same as the work done on the engine under ideal circumstances, this is usually not the case. Usually, the mechanic would have to do more work, because friction will take some of the energy he puts into the lifting and turn it into heat, noise, and other forms of energy that are of no use in the task at hand. The efficiency of the machine tells us how much of the mechanic's work was actually used in lifting the engine

$$\text{efficiency} = \frac{\text{A.M.A.}}{\text{I.M.A.}} \times 100\%$$

If a machine were perfect and had no mass and no friction, it would have an efficiency of 100%. It will be your task in this laboratory experiment to determine the efficiency of various simple machines.

Investigation 1

Objective

To determine the efficiency of a simple machine.

Procedure

1. Build the following machine as indicated using the equipment at hand. The pulleys, posts, and rods may be connected in any way that works, so feel free to be a creative problem solver. Make sure that the 500 gram load can be lifted at least 30 cm off the table.

2. Hang a 500 gram mass as the load to be lifted by the machine. Measure this mass precisely:

$$m_{\text{load}} = \underline{\hspace{2cm}} \text{ grams}$$

3. Connect a mass hanger to the string where the diagram shows you are to pull down. This represents the force that you will be using to try to lift the load.

4. By adding slotted masses to the hanger determine:

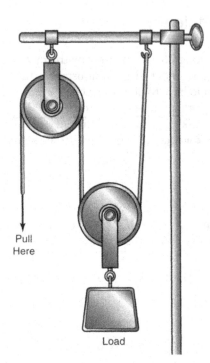

Machine A

   a. What is the least amount of mass that will lift the 500 gram load without help? Measure the mass on a scale (including the mass of the hanger), and record your measurement below:

$$m_{\text{max}} = \underline{\hspace{2cm}} \text{ grams}$$

   b. What is the smallest mass that will lift the 500 gram load if you give it occasional nudges when it gets stuck? Since the mass that's doing the lifting will be coming down, a fair way to nudge it is to tap it upwards to get it unstuck. Again, measure this mass on a scale, including the mass of the hanger:

$$m_{\text{min}} = \underline{\hspace{2cm}} \text{ grams}$$

There are two types of friction that affect our determination of how much the machine multiplies force. There's static friction, which we have to beat to get the motion started. That's what we did in 4a. Once it's moving, though, we only have to fight kinetic friction, which is usually weaker.

   c. Is the mass in 4b smaller than the mass in 4a? \underline{\hspace{2cm}}

Determining the I.M.A.

5. By pulling on the string, raise the 500 gram load as high as you can. Record:

   a. How high the load is lifted: d_{load} = _____ cm

   b. How much you had to pull: d_{lift} = _____ cm

Interpretation

1. What is the I.M.A. of this machine? Use the practical definition of I.M.A., along with the lengths you measured in 5a and 5b.

2. Here's a multiple choice question. Your measurements in 5a and 5b may be a little off. Actually, the true I.M.A. of this particular machine is supposed to be: (Circle one)

   a. 1/2

   b. 2

   c. 4

   d. 6

3. What is the minimum A.M.A. of this machine? Use the mass from 4a:

$$\text{A.M.A.}_{min} = \frac{F_{load}}{F_{pull}} = \frac{m_{load}g}{m_{max}g} = \underline{\hspace{1cm}}$$

4. What is the maximum A.M.A. of this machine? Use the mass from 4b:

$$\text{A.M.A.}_{max} = \frac{F_{load}}{F_{pull}} = \frac{m_{load}g}{m_{min}g} = \underline{\hspace{1cm}}$$

Because of the friction in the pulleys, there is some uncertainty over what the mechanical advantage will be when the pulley system is used in a given situation. In this case, scientists would record the A.M.A. as being within this range.

5. Calculate the efficiency of the machine. Do it once using A.M.A.$_{max}$ and once using A.M.A.$_{min}$. In both cases, use the value of I.M.A. that you chose in question 2. Express your answer as a range:

The efficiency is between   _____% and _____%

Investigation 2

Objective

To determine the efficiency of a different machine following the same procedure as in Investigation 1.

Procedure

We will repeat the procedure of Investigation 1, for the following machine.

1. Build the following machine as indicated using the equipment at hand. Make sure that the 500 gram load can be lifted at least 30 cm off the table.

2. Hang a 500 gram mass as the load to be lifted by the machine. Measure this mass precisely:

$$m_{load} = \text{_____ grams}$$

3. Connect a mass hanger to the string where the diagram shows you are to pull down. This represents the force that you will be using to try to lift the load.

4. By adding slotted masses to the hanger determine:

   a. What is the least amount of mass that will lift the 500 gram load without help? Measure the mass on a scale (including the mass of the hanger), and record your measurement below:

   $$m_{max} = \text{_____ grams}$$

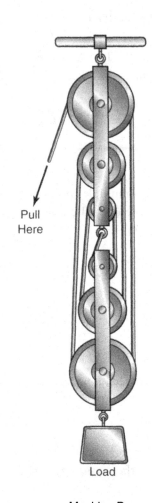

Pull Here

Load

Machine B

   b. What is the smallest mass that will lift the 500 gram load if you give it occasional nudges when it gets stuck? Since the mass that's doing the lifting will be coming down, a fair way to nudge it is to tap it upward to get it unstuck. Again, measure this mass on a scale, including the mass of the hanger:

   $$m_{min} = \text{_____ grams}$$

   c. Is the mass in a larger than the mass in b?   _____

5. By pulling on the string, raise the 500 gram load as high as you can. Record:

    a. How high the load is lifted: d_{load} = _____ cm

    b. How much you had to pull: d_{lift} = _____ cm

Interpretation

1. What is the I.M.A. of this machine? Use the practical definition of I.M.A., along with the lengths you measured in 5a and 5b.

2. Actually, the true I.M.A. of this machine is supposed to be: (Circle one):

    a. 1/3

    b. 2

    c. 4

    d. 6

3. What is the minimum A.M.A. of this machine? Use the mass from 4a:

$$\text{A.M.A.}_{min} = \frac{F_{load}}{F_{pull}} = \frac{m_{load}g}{m_{max}g} = \underline{\hspace{1.5cm}}$$

4. What is the maximum A.M.A. of this machine? Use the mass from 4b:

$$\text{A.M.A.}_{max} = \frac{F_{load}}{F_{pull}} = \frac{m_{load}g}{m_{min}g} = \underline{\hspace{1.5cm}}$$

5. Calculate the efficiency of the machine. Do it once using A.M.A.$_{max}$ and once using A.M.A.$_{min}$. For both cases, use the true I.M.A. you chose from Question 2 above. Express your answer as a range:

The efficiency is between   _____% and _____%

Investigation 3

Objective

To determine the efficiency of a different machine, following the same procedure as in Investigations 1 and 2.

Procedure

We will repeat the previous procedure for the following machine.

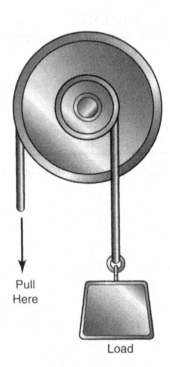

Pull
Here

Load

Machine C

1. Build the following machine as indicated using the equipment at hand. Make sure that the 500 gram load can be lifted at least 30 cm off the table. Note that there are two strings involved: one that unreels as the other one gets reeled up.

2. Hang a 500 gram mass as the load to be lifted by the machine. Measure this mass precisely:

$$m_{load} = \text{_____ grams}$$

3. Connect a mass hanger to the string where the diagram shows you are to pull down. This represents the force that you will be using to try to lift the load.

4. By adding slotted masses to the hanger determine:

a. What is the least amount of mass that will lift the 500 gram load without help? Measure the mass on a scale (including the mass of the hanger), and record your measurement below:

$$m_{max} = \text{_____ grams}$$

b. What is the smallest mass that will lift the 500 gram load if you give it occasional nudges when it gets stuck? Since the mass that's doing the lifting will be coming down, a fair way to nudge it is to tap it upwards to get it unstuck. Again, measure this mass on a scale, including the mass of the hanger:

$$m_{min} = \text{_____ grams}$$

5. By pulling on the string, raise the 500 gram load as high as you can. Record:

a. How high the load is lifted: $d_{load} = \text{_____ cm}$

b. How much you had to pull: $d_{lift} = \text{_____ cm}$

Interpretation

1. What is the I.M.A. of this machine? Use the practical definition of I.M.A., along with the lengths you measured in 5a and 5b.

2. What is the minimum A.M.A. of this machine? Use the mass from 4a:

$$\text{A.M.A.}_{min} = \frac{F_{load}}{F_{pull}} = \frac{m_{load}g}{m_{max}g} = \underline{\hspace{1.5cm}}$$

3. What is the maximum A.M.A. of this machine? Use the mass from 4b:

$$\text{A.M.A.}_{max} = \frac{F_{load}}{F_{pull}} = \frac{m_{load}g}{m_{min}g} = \underline{\hspace{1.5cm}}$$

4. Calculate the efficiency of the machine. Do it once using A.M.A.$_{max}$ and once using A.M.A.$_{min}$. Express your answer as a range:

The efficiency is between _____% and _____%

Investigation 4

Objective

To determine the efficiency of a different machine.

Procedure

We will repeat the previous procedure for the following machines.

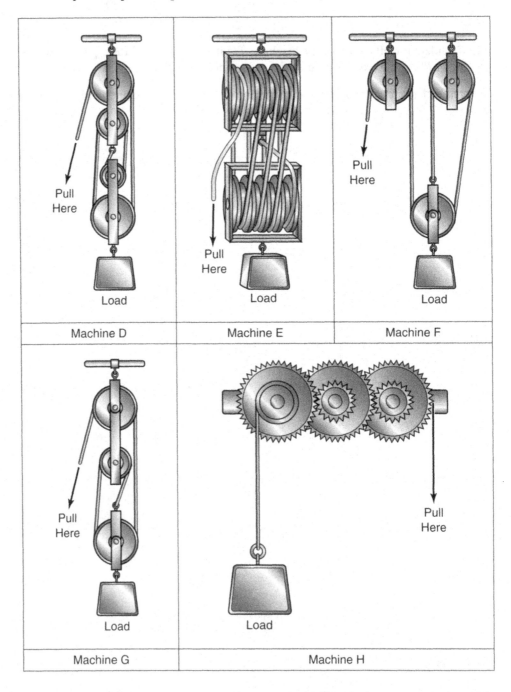

1. Choose any one of the machines from the diagram. Build it as indicated using the equipment at hand. Make sure that the 500 gram load can be lifted at least 30 cm off the table.

2. Hang a 500 gram mass as the load to be lifted by the machine. Measure this mass precisely:

$$m_{load} = \underline{\hspace{2cm}} \text{ grams}$$

3. Connect a mass hanger to the string where the diagram shows you are to pull down. This represents the force that you will be using to try to lift the load.

4. By adding slotted masses to the hanger determine:

   a. What is the least amount of mass that will lift the 500 gram load without help? Measure the mass on a scale (including the mass of the hanger), and record your measurement below:

$$m_{max} = \underline{\hspace{2cm}} \text{ grams}$$

   b. What is the smallest mass that will lift the 500 gram load if you give it occasional nudges when it gets stuck? Since the mass that's doing the lifting will be coming down, a fair way to nudge it is to tap it upwards to get it unstuck. Again, measure this mass on a scale, including the mass of the hanger:

$$m_{min} = \underline{\hspace{2cm}} \text{ grams}$$

5. By pulling on the string, raise the 500 gram load as high as you can. Record:

   a. How high the load is lifted: $d_{load} = \underline{\hspace{2cm}} \text{ cm}$

   b. How much you had to pull: $d_{lift} = \underline{\hspace{2cm}} \text{ cm}$

Interpretation

1. What is the I.M.A. of this machine? Use the practical definition of I.M.A., along with the lengths you measured in 5a and 5b.

2. Actually, the true I.M.A. of these machines should be one of the following (Choose one):

   a. 3

   b. 4

   c. 8

   d. Other: \underline{\hspace{1.5cm}}

3. What is the minimum A.M.A. of this machine? Use the mass from 4a:

$$\text{A.M.A.}_{min} = \frac{F_{load}}{F_{pull}} = \frac{m_{load}g}{m_{max}g} = \underline{\hspace{2cm}}$$

4. What is the maximum A.M.A. of this machine? Use the mass from 4b:

$$\text{A.M.A.}_{max} = \frac{F_{load}}{F_{pull}} = \frac{m_{load}g}{m_{min}g} = \underline{\hspace{2cm}}$$

5. Calculate the efficiency of the machine. Do it once using A.M.A.$_{max}$ and once using A.M.A.$_{min}$. Express your answer as a range:

The efficiency is between   _____% and _____%

Report Sheet

Experiment 10

Name: _____

Date: _____   Section: _____

1. From your investigations of each machine, collect the following information to fill in the table below. Be sure to specify which machine you used in Investigation 4.

Machine	I.M.A.	A.M.A. (range)	Efficiency (range)
A			
B			
C			
Your choice:			

2. What factors contribute in making these machines less than 100% efficient?

3. Ideally, the I.M.A. of machine B should be 6. How could you have predicted that, based on the diagrams?

4. Which of these machines would make it easiest to lift a 1-ton car? Justify your choice.

5. Let's say you do 100 joules of work in lifting a load using machine B. Does that mean you gave the load 100 joules of gravitational potential energy? Would it crash to the ground with 100 joules of kinetic energy if you dropped it?

Extra Credit

1. Sometimes a machine can be used backward, to divide a force. In that case, you would have to exert a larger force, but you would get more movement on the other end. Give a real life example of a machine that is used backwards, and explain its usefulness.

2. It is said that the pyramids of Egypt were built by pushing slabs of rock up long ramps. Otherwise, the slabs were too heavy to lift straight up. How can we say that the ramp is a machine?

How do the volumes and absolute temperatures of a gas vary?

Volume and Temperature Relationships of Gases

Background Concepts

There are three generally recognized states of matter: solids, liquids, and gases. Of the three, gases are defined as having neither a fixed shape nor volume. Gases are the least conspicuous state of matter, especially when they are both odorless and colorless. A gas sample may be composed of either a pure substance, such as helium, oxygen, or carbon dioxide, or it may be a mixture of several pure substances. The most familiar gas mixture is air, which is approximately 78% nitrogen, N_2, 21% oxygen, O_2, and 1% other gases. The "other gases" can include water vapor, carbon dioxide, inert gases, and a variety of other gases, dependent on the location.

Little investigation of the gaseous state was done until after 1654 when, through the development and use of an air pump, the German physicist Otto von Guericke proved that a vacuum could be established and maintained. Subsequent investigators such as Boyle, Charles, and Gay-Lussac did much to add to our present knowledge of the behavior of gases under conditions of varying temperatures, pressures, and volumes. An understanding of the behavior of gases also paved the way for the later development of steam and internal combustion engines.

Through careful experimentation and our understanding of the kinetic molecular theory, it is now known that the behavior of all gases at low pressures and high temperatures are essentially the same. The mathematical relationship that defines an "ideal" gaseous system is called an equation of state and is given by the equation:

$$PV = nRT$$

In the ideal gas equation, the pressure, P, volume, V, the number of moles of gas, n, and its absolute temperature, T, are related by the universal gas constant, R. The value of R, therefore, varies according to the units used, but some of the more commonly used values are 8.314 J/mol · K; 0.08206 liters atm/mole · K; and 1.987 cal/mole · K.

One of the common ways to experimentally determine the effect of one of the variables on another is to hold all the other variables constant. For example, if a closed container is used, then the amount of gas in the sample remains constant. In this situation, n and R are both constant and their product would also be a constant, k_n, for example. The relationship between the remaining variables could therefore be rewritten as:

$$\frac{PV}{T} = k_n$$

In another example, if both the amount of gas in a sample ($n = k$), the temperature of the sample ($T = k$) remain constant, and R is constant ($R = k$), then the relationship between the remaining variables could be rewritten as:

$$PV = nRT \quad \text{or} \quad PV = k_{nT}$$

This relationship is known as Boyle's law and states that the volume and absolute pressure of a gas are directly proportional, if all other variables are held constant.

In this experiment, we will examine yet another relationship between the volume and absolute temperature of a gas sample, while holding all the other variables constant. From the experimentally determined relationship, we will be able to write a mathematical expression for Charles' law.

▌ Investigation 1

Objective

To determine the relationship between the change in the length of an air column and absolute temperature, measured at constant pressure.

Procedure

1. Select a plastic dropper.

2. Place the tip of the dropper under water and slowly release 7 or 8 air bubbles, by gently squeezing the bulb. Once the last bubble is expelled, release the bulb and allow it to draw in a column of water.

3. Remove the dropper from the water and squeeze the bulb gently to expel enough water so that the column of water remaining is only approximately 1 cm in length. Release the bulb and allow the water to move into the tube as shown in Figure 11-1. Repeat steps 2 and 3, if necessary, by expelling one more or less bubbles until you have a 1 cm column of water that is positioned approximately 1 cm from the bulb, while laying on the counter top.

> **Note:** It is important that you not be touching the bulb while attempting to duplicate Figure 11-1. You will note that even brief contact of your fingers with the bulb will warm the trapped air and cause the water column to move outward from the bulb. From this point on in the investigation, *only* handle the dropper by the tip of the tube.

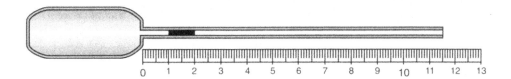

Figure 11-1: A 1 cm column of water located approximately 1 cm from the bulb.

4. Use a metric rule to measure the distance from the bulb to the closest edge of the column of water. Record this length in Table 11-1.

5. Lay a thermometer next to the bulb on the counter top. Read and record the temperature in Table 11-1, and assume that it is the same as the temperature of the air in the bulb.

6. Prepare a water temperature bath by use a small (100 or 150 mL) beaker and enough water to submerge the bulb of the dropper.

7. Place a second small beaker of water on a hot plate and warm it to about 50°C.

8. Adjust the temperature of the water bath by adding warm water a dropper full at a time, until it is one or two degrees warmer than the temperature recorded in step 5.

---

9. While holding the dropper by the tube tip, submerge the bulb into the water bath. Wait a minute before using a metric rule, as shown in Figure 11-2, to measure and record in Table 11-1 the distance from the bulb to the closest edge of the column of water.

10. Read and record in Table 11-1, the temperature of the water in the beaker immediately after making the length measurement.

11. Repeat steps 8 and 9 at least five times, or until the volume of gas becomes large enough that it appears it will pushed all the water out of the tube. Record the distance from the bulb to the closest edge of the column of water and the temperature immediately after making each measurement in Table 11-1.

Figure 11-2

Trial Number	Distance		Temperature	
	Bulb to Edge of Water (mm)	Change in Distance to Edge of Water (mm)	°C	K
Start		0.0 mm		
0				
1				
2				
3				
4				
5				
6				

Table 11-1: Length of air column from bulb and temperature data.

Interpretation

I. Graphing:

> **Note:** You may prepare the following graph manually. Although there are other spreadsheets that may be used for this part of the experiment, the directions that follow are specific for Microsoft's Excel® software.

1. Complete Table 11-1, by calculating the *change* in the distance from the starting position to the edge of the water column and converting each temperature to absolute temperature ($K = °C + 273$).

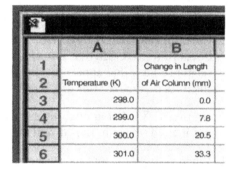

Figure 11-3

2. Plot the data in Table 11-1 on a graph using the following steps.

3. On your computer, start Microsoft Excel®. You should have an empty worksheet on your screen.

4. In the first column (Column A), enter the absolute temperatures, K, as shown in Figure 11-3. This will correspond to the *x*-component of the data points to be plotted.

5. In the second column (Column B) enter the *change* in the length of the air column from the starting point that corresponds to each of these temperatures. These values will correspond to the *y*-components of the points to be plotted.

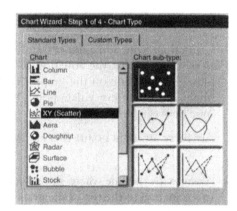

Figure 11-4

6. By clicking and dragging the mouse over the two columns, select both columns.

7. Under the "Insert" menu, select "Insert Chart".

8. As shown in Figure 11-4, choose an "XY (Scatter Plot)" as your chart type. This will produce a graph whose points get their *x* and *y* coordinates from your data.

9. In the NEXT dialog box, make sure that your data series is to be interpreted in COLUMNS. The data range that's listed should correspond to the cells you selected in step 6. Normally it also shows you this with a blinking line drawn around your data.

---

10. In the NEXT dialog box, (see Figure 11-5) you will have the option of tailoring your graph. Click on the TITLES tab, if it isn't already selected. This gives you the choice of giv-

Figure 11-5

ing names to the graph, the x-axis, and the y-axis. Since we are putting distance on the y-axis, enter "Length of Air Column (mm)" as the title for the y-axis. Similarly, put "Temperature (K)" as the title of the x-axis. You should also give your graph a title, like "Change in Length of Air Column vs. Absolute Temperature."

11. Click on the LEGEND tab of this dialog box. Since we are only plotting one set of data, we don't need a legend to help us distinguish one set of points from another. Click on the check box so that the Legend is not shown.

12. Click on the GRIDLINES tab of this dialog box. Click on the different check boxes to try out different sets of grid lines. These are the horizontal and vertical reference lines that appear on the graph to help you later read the values. Select the ones that make your graph readable, without making it cluttered.

13. In the NEXT dialog box, the computer asks where to put the chart. Choose to put it as an object in the current sheet.

14. When your graph appears, click on its corner, stretch it, and move it so that it fits beside your data.

II. Analysis of Data

1. While you have the graph on your screen, select "Add Trendline..." You can find this option in the CHART menu.

2. Using the menu, add a *linear* trendline. Set its *options* so that it has an *intercept set to zero* and so that the equation of the trendline appears on the chart. This will plot a straight line on the graph that best agrees with the data points. The line might not touch every data point, but it should appear as a reasonable compromise between all of the points.

   Are there any data points that are far away from the line?

3. Along with the line, there should have appeared an equation for the line. Move that equation to the top of the graph, so that it can be seen more clearly. Write the equation below:

   $y =$ _____

4. Since the computer doesn't really know what your data represents, it uses the symbols "y" and "x" in the equation to represent the quantities that you're plotting on the vertical axis and the horizontal axis, respectively. You have plotted change in length of air column on the vertical or y-axis and absolute temperature on the horizontal axis or x-axis. Re-write the equation in step 3 using ΔL for the change in length term and T, for the temperature term.

5. Print out a copy of your graph, along with the data beside it. Attach it to your final report sheet.

Investigation 2

Objective

To determine the relationship between the volume of air trapped in the dropper and the length of the air column in the tube.

Procedure

1. Using the same dropper, squeeze the bulb several times to remove all of the water in the tube. Use a dry tissue or paper towel to remove any water that may be clinging to the tube or tip of the dropper.

2. Measure, to the nearest ± 1 mm, the length of the tube from the base of the bulb to its tip. Record this length (cm) in Table 11-2.

3. Weigh the dry dropper to the nearest ± 0.01 grams and record its mass in Table 11-2.

4. Using a small beaker of water at room temperature, repeatedly squeeze the bulb, until you have completely filled the dropper bulb and tube. Use your thermometer to determine the temperature of the water and record it in Table 11.2.

5. Dry the outside of the dropper, weigh the dropper and water to the nearest ± 0.01 grams, and record the total mass in Table 11-2.

6. Devise a method to adjust the amount of water in the dropper so that the bulb is *filled*, but there is no water in the tube.

7. Weigh the dropper with the filled bulb to the nearest ± 0.01 grams and record the mass in Table 11-2.

	Empty Dropper	Filled Dropper	Bulb Filled Dropper
Length of Dropper Tube	cm	****	****
Total Mass	g	g	g
Temperature of H_2O	****	°C	°C
Density of H_2O	****	g/mL	g/mL
Mass of H_2O, only	****	g	g
Volume of H_2O	****	mL	mL

Table 11–2: Mass of dropper and water data.

Interpretation

1. Use Table 11-3 to determine the density of water at the temperature recorded. Do the calculations necessary to complete Table 11-2.

2. What is the volume of just the dropper bulb? Express this answer in milliliters, mL, cubic centimeters, cm^3, and cubic millimeters, mm^3. (Show your work below.)

Density of Water			
t, °C	*d,* gm/mL	*t,* °C	*d,* gm/mL
0	0.99987	40	0.99224
3.98	1.00000	45	0.99025
5	0.99999	50	0.98807
10	0.99973	55	0.98573
15	0.99913	60	0.98324
18	0.99862	65	0.98059
20	0.99823	70	0.97781
25	0.99707	75	0.97489
30	0.99567	80	0.97183
35	0.99406	85	0.96865
38	0.99299	90	0.96534
		95	0.96192
		100	0.95838

Weast, Robert C., Editor, Handbook of Chemistry and Physics, CRC Press, Cleveland, Ohio, 1975 p. F-11.

Table 11–3

3. Determine the volume of the dropper tube by finding the difference in water volume between the "Filled Dropper" and the "Bulb Filled Dropper" columns. Express this answer in milliliters, mL, cubic centimeters, cm^3, and cubic millimeters, mm^3. (Show your work below.)

4. Use the volume of water difference, expressed in mm^3 and the length of the tube expressed in millimeters, mm, to determine a volume to length of tube ratio, mm^3/mm, for your dropper. (Show your work below.)

Investigation 3

Objective

To determine the relationship between the volume and absolute temperature of an air sample, at constant pressure.

Procedure

1. Using the information recorded in Table 11-1, the volume of the bulb calculated in Interpretation step 2 of Investigation 2, and the mm^3/mm ratio calculated in Interpretation step 3 of Investigation 2, complete Table 11-4.

Trial Number	Distance from Bulb to Edge of H₂O (mm)	Volume Air in Tube, (mm × mm³/mm)	Volume Air in Bulb (constant)	Total Volume of Air (mm³)	Temperature (K)
Start					
1					
2					
3					
4					
5					
6					

Table 11–4: Volume of air trapped vs. absolute temperature data.

Interpretation

I. Graphing:

We will again use Excel® to prepare a graph of the data.

1. Open a new Excel® file on your computer. Start with the first row and move to the right. Type headings of: "Volume" and "Temperature." (One word in each cell.)

2. Place the information calculated in the last two columns of Table 11-4 below the appropriate headings.

3. Using the same steps as used in the Interpretation section of Investigation 1, prepare a graph of the total air volume vs. absolute temperature data.

II. Analysis of Data

1. How does the slope of this graph compare to the one prepared in Investigation 1? (Write a brief explanation for any similarities or differences found.)

2. Print out a copy of the data and graph. Attach it to your final report.

3. Complete Table 11-5 to determine if a constant relationship, k_p, exists between the volume of a gas, V, and its absolute temperature, T, at constant pressure.

Trial Number	Volume, (mm³)	Temperature, (K)	Constant (k_p)	
			$V \times T$	V/T
Start				
1				
2				
3				
4				
5				
6				

Table 11–5: Search for a relationship between V and T at constant pressure.

4. Which of the columns, $V \times T$ or V/T, produces the most constant relationship, k_p?

5. Write the mathematical formula relating volume, V, absolute temperature, T, and a constant, k_p.

6. Is the mathematical relationship that exists between the variables an inverse or a direct proportionality?

7. Starting with the equation of state, $PV = nRT$, rearrange the terms and determine the ones that are included in the constant, k_p.

Report Sheet

Name: _____

Experiment 11

Date: _____ **Section:** _____

1. In Investigation 1, what was the equation for the line, expressed in length, ΔL, and absolute temperature, T.

2. Attach a copy of your Investigation 1 data and graph to this report.

3. What is the volume of the dropper bulb, expressed in milliliters, mL, cubic centimeters, cm^3, and cubic millimeters, mm^3, as determined in Investigation 2.

4. What is the calculated volume to length of tube ratio (mm^3/mm) calculated for the dropper in Investigation 2?

5. Transfer the data from Table 11-4 to complete the following table.

Trial Number	Distance from Bulb to Edge of H_2O (mm)	Volume Air in Tube, (mm × mm^3/mm)	Volume Air in Bulb (constant)	Total Volume of Air (mm^3)	Temperature (K)
Start					
1					
2					
3					
4					
5					
6					

6. Attach a copy of your Investigation 3 data and graph to this report.

7. What is the mathematical formula relating volume, V, absolute temperature, T, and the constant, k_p found in Investigation 3?

Extra Credit

Using the temperature, °C, and volume of air trapped, mm^3, for each of the trials recorded in Tables 11-1 and 11-4, prepare a spreadsheet to graph "Volume of Air vs. Temperature, °C." Determine and use the slope of the line to predict the temperature at which the volume of air would theoretically equal zero. What is the percent error and significance of this temperature? Explain your answers. (Attach your graph and calculations to this report.)

What do a tank of gasoline and a boulder on a cliff have in common?

Energy

Background Concepts

Energy is an abstract, intangible quantity that physical objects possess. Sometimes energy can be seen as the ability to do damage. For example, a truck speeding down the highway and a boulder perched on a cliff both possess energy that can do damage. The truck has kinetic energy by virtue of its speed and mass. It can do a lot of damage to your car if you get in its way. The boulder, on the other hand, has gravitational potential energy by virtue of its height. The boulder can also do a lot of damage to your car if it were to fall on it.

Using the formula, $K = \frac{1}{2} m v^2$ it is possible to calculate the kinetic energy, in joules, of the truck. In the formula, the "m" is for the mass of the truck and "v" is its velocity. The gravitational potential energy can be found by using the formula, $U_{grav} = m g h$. In the formula, "m" is the mass of the boulder, "g" is the gravitational field ($9.8 \text{ N/kg} = 9.8 \text{ m/s}^2$), and "$h$" is its elevation above some reference point, such as the ground.

In this way, a 1,470 kg truck, traveling at 20 m/s has the same amount of energy (2.94×10^5 J) as a 300 kg boulder, perched on a cliff 100 meters above the ground (2.94×10^5 J). Both have the same capacity to do damage to your car: one by crashing into it, the other by falling on it.

Conversion and Conversation of Energy

When a boulder falls from a cliff, it rapidly loses height, but at the same time it is rapidly gaining speed. Its gravitational potential energy is decreasing, but its kinetic energy is increasing. If you calculate and total these two energies every split-second until the boulder reaches the ground, you will find that the totals are always equal. Put another way, the boulder has a constant number of joules of energy. Near the top, most of the joules are in the form of gravitational potential energy. By the time the boulder is near the bottom, most of the joules have been converted into kinetic energy. In this way, gravitational potential energy is converted into kinetic energy, so it's not surprising

---

that the boulder can inflict as much damage as a moving truck. It will have as much kinetic energy as the truck, when it hits your car.

There are also other forms of energy. Gunpowder is a stored form of chemical energy. This energy can be converted into the kinetic energy of a bullet when the gun is fired. A stretched rubber band on a slingshot has elastic potential energy. The hot gases used in steam engines have thermal energy; some of this energy is converted into the kinetic energy of the train. Your body contains chemical energy, in the form of glucose and other carbohydrate molecules. Your body burns these fuel molecules turning some of it into the kinetic energy of your heart, lungs, and limbs.

Energy can be converted from one form to another, but the total amount of energy always remains the same. Suppose, for instance, that when the boulder falls from the cliff its fall is slightly slowed by air resistance. That is, as it rubs against the molecules in the air, they are slightly warmed. Because the air has gained thermal energy, the boulder must have lost an equal amount of kinetic energy. You will find, however, that the boulder does not have the same total kinetic energy at the bottom as it did gravitational potential energy at the top. Part of the gravitational potential energy has been converted into the kinetic energy of the boulder and a part of it has been converted into thermal energy. A careful analysis still shows that the total ($K + U_{grav} +$ thermal energy of air) remains the same. Energy can never be created or destroyed. Accounting for energy is in some ways like accounting for money – when you lose it from your account, it can be traced to another bank account.

Power

It is common to use a machine to transform energy from one form to another. For example, the engine in a car transforms the chemical energy in gasoline into the kinetic energy of the car and some heat. For a race car, the engine must perform this transformation quickly. The faster it occurs, the quicker the car can accelerate. What is important here is how many joules of energy can it transform per second. We call that the power of the engine.

$$\text{power} = \frac{\text{energy transformed}}{\text{time}}$$

When 2,000 joules per second is transformed, we say it has a power of 2,000 watts. A 60-watt light bulb can, therefore, transform electric energy into light and heat energy at a rate of 60 joules per second.

Investigation 1

Objectives

To measure the amount of "damage" done to a chalkboard eraser when it is struck by a rolling ball and measure how fast the ball is rolling when it hits the eraser.

To calculate the amount of kinetic energy a ball has when it strikes the eraser.

Procedure

Energy is an abstract quantity, and therefore hard to visualize. For the purposes of this experiment, we will define energy as the "ability to do damage." We now have a concrete way to discuss it, provided we can find a way to quantitatively measure the "damage." So, instead of doing real damage, we will crash balls into a chalkboard eraser and measure how far back it gets pushed. The distance the eraser gets pushed will be considered a measure of the amount of "damage" done.

1. Place a chalkboard eraser on the floor as shown in Figure 12-1 and tape a meter stick on either side of it. The meter sticks should be close enough to the eraser to prevent it from turning, but they should not squeeze it and prevent it from sliding freely.

2. Set a ramp as shown in Figure 12-2 so one end reaches the floor, a bit away from the meter sticks. It should be positioned so you can roll a ball down the ramp and send it crashing into the eraser.

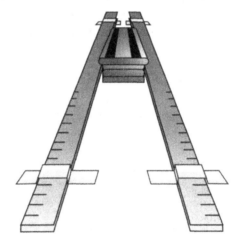

Figure 12-1

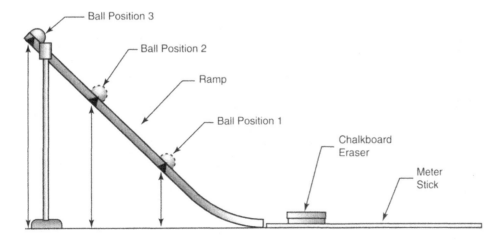

Figure 12-2

3. Measure the height of the other end of the ramp above the floor. Mark a starting position for the ball corresponding to:

 — 1/3 of the way up the ramp

 — 2/3 of the way up the ramp

 — at the top of the ramp

4. Measure and record in Table 12-1 the vertical height of each starting point above the floor.

5. Starting at the lowest position, roll the ball down the ramp and measure the distance the eraser slides. Repeat the measurement three times at each position. Record your data in Table 12-1.

6. Using a stopwatch, measure the velocity of the ball as it comes off the ramp for each of the three starting heights. To do this, lay two meter sticks end-to-end. Roll the ball down the ramp from each starting height. Time how long it takes the ball to roll from one end of the 2 meter sticks, all the way to the other end. Hint: It is sometimes useful to use a third meter stick to form a "V" to help keep the ball "in." Record your measurements in Table 12-1.

7. Determine the mass of the ball in grams and convert it into kilograms. Record this information in Table 12-1.

Trial	Recoil distance of eraser	Time, t, for ball to roll 2 meters	Velocity of ball (= 2 meters/t)	Kinetic Energy of ball (1/2 $m\,v^2$)
Position 1 1/3 of the vertical height _____cm	Trial 1: _____cm Trial 2: _____cm Trial 3: _____cm Avg: _____cm	Trial 1: _____s Trial 2: _____s Trial 3: _____s Avg: _____s	 _____m/s	 _____ joules
Position 2 2/3 of the vertical height _____cm	Trial 1: _____cm Trial 2: _____cm Trial 3: _____cm Avg: _____cm	Trial 1: _____s Trial 2: _____s Trial 3: _____s Avg: _____s	 _____m/s	 _____ joules
Position 3 Top of ramp _____cm	Trial 1: _____cm Trial 2: _____cm Trial 3: _____cm Avg: _____cm	Trial 1: _____s Trial 2: _____s Trial 3: _____s Avg: _____s	 _____m/s	 _____ joules
Mass of ball	_____g	_____kg		

Table 12–1: Investigation 1 data.

Interpretation

1. Based on the three measurements taken, calculate and record in Table 12-1 the average velocity, v, of the balls at each of the different heights.

$$(v = 2 \text{ meters}/\text{time})$$

2. Using the average velocity, v, calculate and record in Table 12-1 the average kinetic energy, $K = \frac{1}{2} mv^2$ of the balls for each of the three heights.

3. Based on your data, do the balls that starts at the highest point (three times higher) do approximately three times as much "damage" as the balls starting at the lowest height? Justify your answer with values.

4. Do the balls that start twice as high as the first balls do twice as much damage?

5. It might be reasonable to expect that the balls that start three times as high as the first balls should be going three times as fast as the first balls. Is this true? Should it be true? Justify your answer by using your calculated values.

6. Do the balls that start at the top of the ramp have three times as much kinetic energy as the ones starting at the lowest position?

Investigation 2

Objectives

To measure the length of time it takes to heat a beaker of water, using an immersion heater.

To compare the amount of thermal energy received by the water to the amount of energy drawn by the heater.

Safety Requirements

- Always wear splash proof safety goggles.
- Do not touch hot objects.
- Use beaker tongs or towels when handling hot glassware.

Procedure

1. Obtain an immersion heater as shown in Figure 12-3. Record its wattage.

   _____ watts

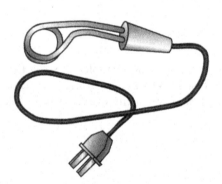

Figure 12-3

2. Select a beaker that the heater will fit into and fill it about $\frac{1}{2}$ full of water at room temperature. Measure and record the following:

   Mass of empty beaker: _____ grams

   Mass of beaker and water: _____ grams

   Mass of water: _____ grams

3. Place the immersion heater in the beaker as shown in Figure 12-4 and turn it on. Using a thermometer, monitor the temperature while gently stirring. Try to keep the thermometer away from the edges of the beaker and the immersion heater.

When the water temperature reaches 30°C, start timing with the stopwatch. Determine and record the length of time required for the temperature to increase from 30°C to 80°C.

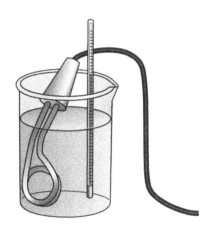

time in seconds: _____ s

Figure 12-4

4. Unplug the immersion heater and gently remove it from the beaker.

> **NOTE:** Do not put the hot immersion heater in cold water.

Interpretation

1. Based on the power rating of the immersion heater, how many joules of electrical energy does it uses each second? (Hint: 1 watt = 1 joule/second).

2. Calculate how much electrical energy was used by the immersion heater during the time the water was heated from 30°C to 80°C.

power × time = _____ joules

3. The specific heat of water is 4,186 J/kg · °C. This means that it takes 4,186 joules of energy to raise the temperature of 1 kg of water by 1 degree. Although you did not use a kilogram of water and you did raised the temperature by more than one degree, use this information to calculate the following:

How many kilograms of water did you heat?   _____ kg

During the timing, what was the temperature change in the water?

_____ degrees

Using the following equation, calculate how much energy was transferred to the water as heat:

Heat energy transferred to water =

4,186 J/kg · °C × _____ kg × _____ °C = _____ J

---

4. How much of the electrical energy was transferred to the water as thermal energy? Express this as a percent efficiency:

$$\frac{\text{heat energy absorbed by water}}{\text{electrical energy used}} \times 100\% = \underline{\hspace{1cm}} \%$$

5. Explain what happened to the rest of the electrical energy.

6. The electrical energy used may have been produced at a hydro-electric dam. At such a dam, as shown in Figure 12-5, the flow of water from the high side to the low side of the dam turns the electrical generator.

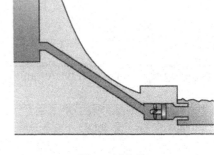

Figure 12-5

Alternatively, the electricity may have been produced by a water-wheel, as shown in Figure 12-6. In either case, what kind of energy was given up by the water and transformed into electrical energy?

7. Sometimes it is more convenient to measure energy in calories. The specific heat of water is also equal to 1 calorie/g · °C. Based on this information, calculate the number of calories absorbed by the water.

Figure 12-6

Investigation 3

Objective

To calculate the energy content of ethanol (ethyl alcohol).

Procedure

1. Obtain an alcohol burner containing ethanol. Determine and record the mass of the burner to the nearest ± 0.01 g.

$$Mass_{begin} = \underline{\hspace{2cm}} grams$$

2. Using a graduated cylinder, measure 100 mL of water and pour it into a 250 mL Erlenmeyer or Florence flask. Support the flask with a lab stand as shown in Figure 12-7. Adjust the height so the bottom of the flask will be about 1cm above the wick of the alcohol burner. Using a thermometer, determine and record the temperature of the water.

$$Temp_{begin} = \underline{\hspace{2cm}} °C$$

3. Place a watch glass over the mouth of the flask. Note the time, light the burner and quickly place it under the flask.

4. Remove and blow out the burner after 5 minutes. Cover the wick to insure that the flame is extinguished.

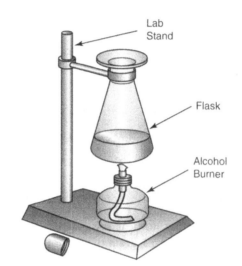

Figure 12-7

5. Carefully remove the hot flask and its contents from the stand. Swirl the water gently, while measuring its temperature. Make sure that the bulb of the thermometer is in the middle of the flask, not resting on the bottom. Record the highest temperature reading.

$$Temp_{final} = \underline{\hspace{2cm}} °C$$

6. Again determine the mass of the alcohol burner to the nearest ± 0.01 gram. Remember to include the mass of the cover, if it was on during your first weighing of the burner.

$$Mass_{final} = \underline{\hspace{2cm}} grams$$

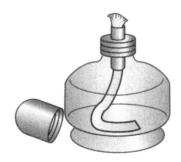

Figure 12-8

Interpretation

1. Based on the mass difference of the burner before and after heating, calculate how much of the ethanol burned.

_____ grams

2. If the specific heat of water is 4,186 J/kg · °C, how many joules should it take to raise the temperature of 100 mL of water by 1 degree Celsius?

(Assume 1 mL = 1 g of water.)

_____ joules

3. Calculate the temperature change of the water during heating.

_____ °C

4. Calculate the number of joules of heat absorbed by the water.

_____ joules

5. Assuming no heat loss, calculate the number of joules of chemical energy produced when 1 gram of ethanol is turned into heat.

_____ joules/gram ethanol

6. In some states, gasoline contains ethanol. As an approximation, let's say that gasoline has the same energy content per gram as you calculated for ethanol above. Calculate the amount of energy that would be stored in 20 gallons of gasoline?

(1 gallon = 3.785 liters and gasoline has a density of 700 grams per liter)

a. volume of gasoline in liters: _____ liters

b. mass of the gasoline: _____ grams

c. total energy: _____ joules

Report Sheet

Experiment 12

Name: _____

Date: _____ Section: _____

1. Using the data and calculations recorded for Investigation 1 in Table 12-1, complete the following:

Trial	Recoil distance of eraser (average)	Velocity of ball (average)	Kinetic Energy of ball
Position 1 vertical height _____ cm	_____ cm	_____ m/s	_____ joules
Position 2 vertical height _____ cm	_____ cm	_____ m/s	_____ joules
Position 3 vertical height _____ cm	_____ cm	_____ m/s	_____ joules

2. Transfer your Investigation 2 findings:

   a. Thermal energy absorbed by water: _____ J

   b. Electrical energy used: _____ J

   c. Efficiency: _____ %

3. Suggest what might have made the efficiency of the immersion heater less than 100%.

4. Transfer your Investigation 3 findings:

   a. Thermal energy absorbed by water: _____ J

   b. Mass of ethanol consumed: _____ g

   c. Energy content of ethanol: _____ J/g

5. A one metric ton (1,000 kg) boulder falls from a 200 m cliff. How much gasoline would it take to have a similar destructive ability? (Hint: see question 6 at the end of Investigation 3.)

Extra Credit

1. Suggest a way in which:

   a. gravitational potential energy can be turned into heat.

   b. chemical energy can be turned into kinetic energy.

   c. electrical energy can be turned into gravitational potential energy.

2. A Btu (British thermal unit) is defined as the energy necessary to raise the temperature of 1 pound of water by 1°F (from 63°F to 64°F). In other words, 1 Btu = 252 calories. Often, a water heater will be labeled 20,000 Btu's, meaning that it can provide 20,000 Btu's per hour. How many Btu's per hour does the immersion heater used in Investigation 2 deliver?

3. An average household uses about 5,000 kilowatt-hours of electrical energy in a year. If the household switched to ethanol, how many kilograms of ethanol would it use per year to provide an equivalent amount of energy?

4. Repeat Investigation 3, using methanol (methyl alcohol) or 1-butanol (n-butyl alcohol) in place of ethanol (ethyl alcohol). Compare the number of carbon atoms and the energy content of the two alcohols.

Do chemical reactions always involve an exchange of energy?

Heat of Reaction

Background Concepts

Thermochemistry is the branch of chemistry concerned with heat exchanges accompanying chemical reactions. The first law of thermodynamics tells us that energy can neither be created nor destroyed, but only changed from one form, such as chemical energy, to another form, such as heat. If heat, Q, flows from a reaction, it is said to be exothermic and the heat of the reaction is assigned a negative number. If heat is absorbed by the reaction, the process is endothermic and the heat of reaction is assigned a positive number.

Heat of reaction is a general term that is frequently replaced by terms that are more descriptive, such as heat of combustion, heat of solution, heat of neutralization. In fact, the heat of reaction represents the difference in the enthalpies, ΔH, of the reaction products and reactants at constant pressure and at a specific temperature, with every substance in a specific physical (solid, liquid, or gas) state. Knowledge of the heat of reaction or enthalpy can aid chemists in determining such characteristics as spontaneity, rate, and completeness.

Germain Henri Hess published a paper on the heats of formation of compounds in 1840, summarizing his thermochemical studies. His conclusions are now called Hess' law, and state that the heat evolved or absorbed in a chemical process is the same whether the process takes place in one or several steps.

When possible, ΔH values are experimentally determined using a device called a calorimeter. Generally, a calorimeter is an isolated chamber designed to measure the heat produced by a reaction. For example, a known amount of pure carbon, C, can be placed in a calorimeter, surrounded by an atmosphere of excess pure oxygen, O_2, ignited, and allowed to burn. The chemical product from this reaction is nearly pure carbon dioxide, $CO_{2\,(g)}$, but a large amount of heat energy is also given off. Reactions of this type are said to be exothermic and ΔH values are assigned a negative value indicating that products contain less entropy than the reactants. The equation for the reaction follows:

$$C_{(s)} + O_{2(g)} \rightarrow CO_{2(g)} \qquad \Delta H = -94{,}050 \text{ cal/mole or } -393.7 \text{ kJ/mole}$$

If the reaction is of the type that absorbs heat energy, such as the thermal decomposition of mercury(II) oxide, $HgO_{(s)}$, then the reaction is said to be endothermic and ΔH has a positive value. The equation for the reaction is:

$$HgO_{(s)} \rightarrow Hg_{(l)} + O_{2(g)} \qquad \Delta H = +21{,}680 \text{ cal or } +90.8 \text{ kJ/mole}$$

When direct calorimetric measurements are not possible or cannot be practically carried out, indirect methods involving Hess' law can be used to determine ΔH values for many other reactions. For example, it is not possible to directly measure the heat involved when carbon burns to carbon monoxide gas, $CO_{(g)}$, in a limited amount of oxygen because some of the $CO_{(g)}$ gets converted into $CO_{2(g)}$. It is, however, possible to obtain a pure sample of carbon monoxide gas and allow it to burn, according to the following equation:

$$CO_{(g)} + \tfrac{1}{2}O_{2(g)} \rightarrow CO_{2(g)} \qquad \Delta H = -67{,}630 \text{ cal or } -283.1 \text{ kJ/mole}$$

Hess' law tells us that it takes the same amount of energy to decompose a compound as is involved in its formation. If its formation is exothermic, then its decomposition will be endothermic, and visa versa. The last equation can therefore be reversed, including the sign on ΔH.

$$CO_{2(g)} \rightarrow CO_{(g)} + \tfrac{1}{2}O_{2(g)} \qquad \Delta H = +67{,}630 \text{ cal or } +283.1 \text{ kJ/mole}$$

Armed with this equation, Hess' law, a knowledge of algebra, and the equation for the burning of pure carbon we now have the necessary information to calculate ΔH for the burning of carbon to $CO_{(g)}$ only.

$$C_{(s)} + O_{2(g)} \rightarrow CO_{2(g)} \qquad \Delta H = -94{,}050 \text{ cal or } -393.7 \text{ kJ/mole}$$

$$CO_{2(g)} \rightarrow CO_{(g)} + \tfrac{1}{2}O_{2(g)} \qquad \Delta H = +67{,}630 \text{ cal or } +283.1 \text{ kJ/mole}$$

$$\overline{C_{(s)} + \tfrac{1}{2}O_{2(g)} \rightarrow CO_{(g)} \qquad \Delta H = -26{,}420 \text{ cal or } -110.6 \text{ kJ/mole}}$$

In the Investigations you will measure the amount of heat, Q, released or absorbed by the reaction of a given amount of reactant in an aqueous medium. From this measurement, the heat of reaction or enthalpy, ΔH, will be calculated for one mole of reactant or product. You will also be asked to predict the heat of reaction, using Hess' law and then experimentally verify the prediction.

Investigation 1

Objectives

To determine if energy is involved when a substance dissolves in water and to determine if energy is involved in an acid-base neutralization reaction.

Safety Requirements

- Always wear splash proof safety goggles.

- Do not touch any chemicals, especially $NaOH_{(s)}$, which is commonly known as lye.

- Wash skin that is exposed to chemicals in cold water. Then, notify the teacher.

- Discard the waste chemicals as instructed by your teacher.

Procedure

NOTE: When an ionic substance dissolves in water (as shown in Figure 13-1), the heat absorbed or evolved is known as the heat of solution. There are two factors involved. First, energy is required (endothermic) to dissociate the ions of the solid. This is known as lattice energy. Second, energy is released (exothermic) by the solvation of the ions by the water molecules. The heat of solution resulting from the dissolving of a substance in water, therefore, is the sum of these opposing factors. When aqueous acids and bases are mixed, their heat of reaction is called the heat of neutralization.

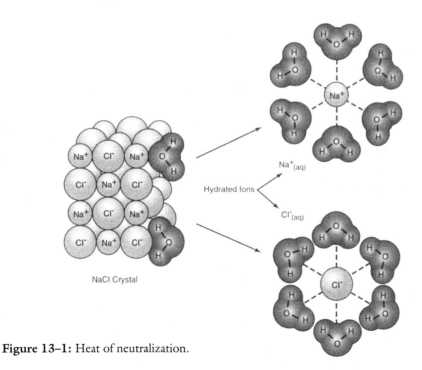

Figure 13–1: Heat of neutralization.

---

1. Using a spatula, put about 2 g of $NaOH_{(s)}$ in a clean dry 150 mL beaker and record its exact mass in Table 13-1.

2. Put 50.0 mL of DI water into your 100 mL graduated cylinder, measure its temperature, and record in Table 13-1.

3. Pour the water into the beaker containing the NaOH and use your thermometer to carefully stir it. What is the temperature when all of the solid has dissolved? Record this temperature in Table 13-1.

4. Set this solution aside for later use in step 7.

5. Repeat steps 1, 2, and 3 using NH_4Cl and DI water. Record the information in Table 13-1.

6. Pour 50.0 mL of 1.0 M $HCl_{(aq)}$ into a clean dry beaker and record its temperature in Table 13-1.

7. Using an ice bath, adjust the temperature of the $NaOH_{(aq)}$ solution set aside in step 4 so that it is the same as the temperature of the $HCl_{(aq)}$.

8. Pour the $HCl_{(aq)}$ into the beaker containing the $NaOH_{(aq)}$ solution. Measure the temperature change and record in Table 13-1.

NaOH Solution Data	
Mass of Beaker, Empty	g
Mass of Beaker and NaOH	g
Mass of NaOH Used	g
Temperature H_2O, Initial	°C
Temperature H_2O, Final	°C
Temperature Change	°C
NH_4Cl Solution Data	
Mass of Beaker, Empty	g
Mass of Beaker and NH_4Cl	g
Mass of NH_4Cl used	g
Temperature H_2O, Initial	°C
Temperature H_2O, Final	°C
Temperature Change	°C
Heat of Neutralization	
Temperature of $HCl_{(aq)}$ Before Mixing	°C
Temperature of $NaOH_{(aq)}$ Before Mixing	°C
Temperature of Mixture	°C
Temperature Change	°C

Table 13–1: Investigation 1 data.

Interpretation

1. Calculate the temperature change that occurred when forming the $NaOH_{(aq)}$ and $NH_4Cl_{(aq)}$ solutions. Record this change in Table 13-1.

2. Is the solution process for dissolving $NaOH_{(s)}$ exothermic or endothermic?

3. Is the lattice energy or the heat of solvation larger for $NaOH_{(s)}$? Explain your answer.

4. Is the solution process for dissolving $NH_4Cl_{(s)}$ exothermic or endothermic?

5. Is the lattice energy or the heat of solvation larger for $NH_4Cl_{(s)}$? Explain your answer.

6. Calculate the temperature change that occurred when the $HCl_{(aq)}$ and $NaOH_{(aq)}$ solutions were mixed. Record the change in Table 13-1.

7. Is the neutralization reaction for $HCl_{(aq)}$ and $NaOH_{(aq)}$ exothermic or endothermic? Explain your answer.

Investigation 2

Objective

To determine the heat of solution, ΔH, for the equation: $NaOH_{(s)} \rightarrow NaOH_{(aq)}$.

Procedure

> **NOTE:** Your instructor may direct you to work with a partner in recording the time-temperature data. All data collected in this investigation should be recorded in Table 13-2.

1. Set up the calorimeter as shown in Figure 13-2.

2. Add 200 mL of DI water to the calorimeter cup. In performing the calculations for this experiment, you may assume that 1.00 mL H_2O = 1.00 g H_2O.

3. Measure and record the temperature of the water in the calorimeter to the nearest $\pm 0.2°C$.

4. Weigh out about 8 g of $NaOH_{(s)}$.

> **NOTE:** Sodium hydroxide pellets are hygroscopic and rapidly gain water from the atmosphere. Do your weighing as quickly as possible to the nearest $\pm 0.01g$, record the mass in Table 13-2, and immediately continue with step 5.

Figure 13–2: Calorimeter setup.

5. Add the weighed $NaOH_{(s)}$ to the water in the calorimeter, noting the time of transfer. The time of transfer is called 0 seconds. Do not record the temperature at 0 seconds.

6. Using the metal stirring wire and holding the thermometer slightly above the bottom of the cup, record the temperature every 30 seconds while mixing. Continue the temperature readings until the temperature is no longer increasing and has held constant for two minutes. When not actually reading the temperature, continue to stir the contents with the wire.

7. Save the $NaOH_{(s)}$ solution prepared in this investigation in a 250 mL beaker for use in Investigation 3.

8. Plot Graph 13-1 using the time-temperature data recorded in Table 13-2, and draw a smooth curve through the points.

9. By means of a straight edge, as shown in Figure 13-3, extrapolate the slope of the cooling curve back to the y-axis. Record the predicted temperature, which represents the highest temperature.

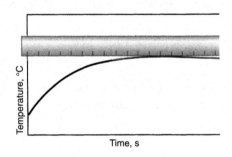

Figure 13–3: Extrapolation of cooling curve to time zero using a straight edge.

Mass of Cold Water		g	
Temperature of Cold Water		°C	
Mass of NaOH$_{(s)}$		g	
Lapse Time After Mixing, *s*	Temperature, °C	Lapse Time After Mixing, *s*	Temperature, °C
0		210	
30		240	
60		270	
90		300	
120		330	
150		360	
180		390	
Predicted Highest Temperature From Extrapolation			°C

Table 13–2: Time-temperature data for dissolving NaOH$_{(s)}$.

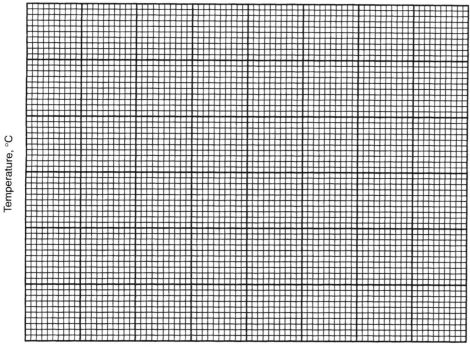

Graph 13–1: Time-temperature plot for solution of NaOH$_{(s)}$.

Interpretation

1. Calculate ΔT for the reaction, using the starting and predicted final water temperatures.

$$\Delta T = \underline{\hspace{3cm}}$$

2. Using 4.186 joule/g · °C, as the specific heat, sp. ht., of the solution, calculate the heat, Q, for the reaction in joules, J, ($Q = $ sp ht $\times$ g $\times \Delta T$). Show your calculation below.

3. How many moles of $NaOH_{(s)}$ were used to produce the calculated amount of heat, Q?

$$\text{moles} = \frac{\underline{\hspace{1.5cm}} \text{ g NaOH}_{(s)}}{40 \text{ g NaOH}_{(s)}/\text{mole}} = \underline{\hspace{1.5cm}} \text{ moles NaOH}_{(s)}$$

4. Calculate ΔH ($\Delta H = Q$/mole) in J/mole for dissolving one mole of $NaOH_{(s)}$. Show your method below.

$$\Delta H = \underline{\hspace{3cm}}$$

5. Refer to the Investigation 1 and assign the proper sign (negative for exothermic and positive for endothermic) and units to the calculated ΔH value.

$$\Delta H = \underline{\hspace{3cm}}$$

Investigation 3

Objective

To determine the heat of neutralization, ΔH, for the following equation:

$$NaOH_{(aq)} + HCl_{(aq)} \rightarrow NaCl_{(aq)} + H_2O.$$

Procedure

1. Use the same procedure followed in Investigation 3 to measure the heat of this reaction. Be sure to dry the calorimeter cup before starting this investigation. Record all data in Table 13-3.

2. Measure out 100 mL of the 1.0 M $HCl_{(aq)}$ solution in a graduated cylinder.

3. Pour the $HCl_{(aq)}$ solution into the clean dry calorimeter cup and record its temperature in Table 13-3.

4. Measure out 100 mL of the $NaOH_{(aq)}$ solution produced in Investigation 2 and adjust its temperature as in Investigation 1, step 7, and record its temperature in Table 13-3.

5. Pour the $NaOH_{(aq)}$ solution into the calorimeter and record the time-temperature data in Table 13-3.

6. Prepare Graph 13-2 from the data collected in Table 13-3.

Volume of $NaOH_{(aq)}$			mL	
Temperature of $NaOH_{(aq)}$			°C	
Volume of $HCl_{(aq)}$			mL	
Temperature of $HCl_{(aq)}$			°C	
Lapse Time After Mixing, _s_	**Temperature, °C**	**Lapse Time After Mixing, _s_**	**Temperature, °C**	
0		210		
30		240		
60		270		
90		300		
120		330		
150		360		
180		390		
Predicted Highest Temperature From Extrapolation			°C	

Table 13–3: Time-temperature data for $NaOH_{(aq)}$ and $HCl_{(aq)}$ neutralization.

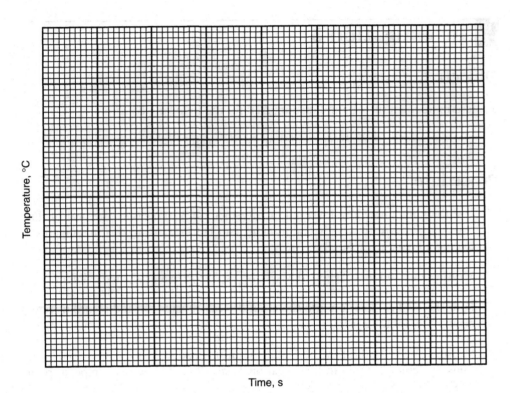

Graph 13–2: Time-temperature plot for $NaOH_{(aq)}$ and $HCl_{(aq)}$ neutralization.

Interpretation

1. Calculate ΔT for the reaction, using the starting and predicted final water temperatures.

$$\Delta T = \underline{\hspace{2cm}}$$

2. Calculate the heat, Q, in both calories and joules for the reaction. Show your calculation below.

3. The moles of $NaOH_{(aq)}$, and $HCl_{(aq)}$ used were approximately equal. Based on the volume and concentration of $HCl_{(aq)}$, how many moles of each were neutralized to produce the calculated amount of heat?

$$molarity_{HCl} = moles_{HCl} / liters_{HCl} \qquad\qquad moles_{HCl} = \underline{\hspace{2cm}}$$

4. Calculate the ΔH in J/mole for the neutralization of one mole of $NaOH_{(aq)}$. Show your method below. Be sure to assign the proper sign (plus or minus) and units.

$$\Delta H = \underline{\hspace{2cm}}$$

Investigation 4

Objective

To discover a method to predict the heat of reaction for the following equation from other experimentally determined heats of reactions:

$$NaOH_{(s)} + HCl_{(aq)} \rightarrow NaCl_{(aq)} + H_2O.$$

Procedure

> **NOTE:** According to Hess' law, chemical equations and their ΔH values can have any mathematical process performed on them that can be performed on any algebraic equation.

1. Enter in the appropriate spaces in Table 13-4 the equations and their ΔH values as determined experimentally in Investigations 2 and 3.

2. Perform any mathematical operation necessary to obtain a chemical equation that is, term for term, the same as the equation stated in the Problem.

	Equation	ΔH value
Investigation 2 Reaction		
Investigation 3 Reaction		
Combined Reaction and Predicted ΔH Value		

Table 13–4: ΔH for $NaOH_{(s)}$ and $HCl_{(aq)}$ reaction.

3. Test your predicted ΔH value by using the same procedures developed in Investigations 2 and 3: reacting 8.00 g of $NaOH_{(s)}$ with 200 mL of 1.0 M $HCl_{(aq)}$. Record your data in Table 13-5. Prepare Graph 13-3 and make necessary calculations.

Mass of NaOH$_{(aq)}$			g
Volume of HCl$_{(aq)}$			mL
Temperature of HCl$_{(aq)}$			°C
Lapse Time After Mixing, *s*	**Temperature, °C**	**Lapse Time After Mixing, *s***	**Temperature, °C**
0		210	
30		240	
60		270	
90		300	
120		330	
150		360	
180		390	
Predicted Highest Temperature From Extrapolation			°C

Table 13–5: Time-temperature data for NaOH$_{(s)}$ and HCl$_{(aq)}$ neutralization.

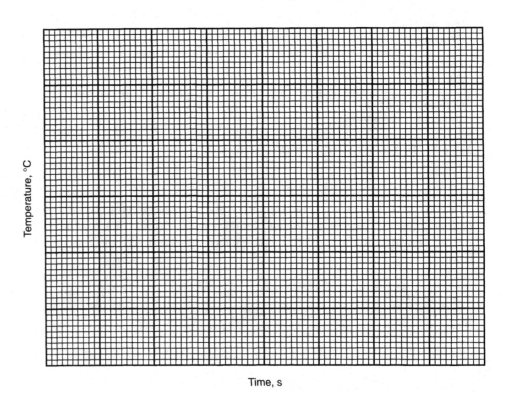

Time, s

Graph 13–3: Time-temperature plot for NaOH$_{(s)}$ and HCl$_{(aq)}$ neutralization.

Interpretation

1. What was your predicted ΔH value for the reaction, expressed in J/mole? Show your method of calculation.

$$\Delta H = \underline{\hspace{2cm}}$$

2. What is your experimentally determined ΔH for the reaction? Show your calculations.

$$\Delta H = \underline{\hspace{2cm}}$$

3. From your above answer, does it appear that the overall heat of reaction is independent of the steps in which the reaction takes place? Explain your answer.

Report Sheet

Name: _____

Experiment 13

Date: _____ Section: _____

1. Complete the following table with the information collected during Investigation 1.

Item	$NaOH_{(s)}$	$NH_4Cl_{(s)}$	$HCl_{(aq)}$ and $NaOH_{(aq)}$
Temperature Change, ΔT			
Exothermic or Endothermic			
Lattice Energy or Heat of Solvation Larger?			

2. Record the equation, temperature change, ΔT, heat, Q, and the ΔH value for Investigations 2 and 3 in the following table.

Item	Equation for the Reaction	ΔT	Heat, Q	ΔH Value, kcal/mole
Investigation 2 Reaction				
Investigation 3 Reaction				

3. Transfer the data from Table 13-4, showing how you made your prediction for Investigation 4. How do the predicted and experimental values for the reaction of $NaOH_{(s)}$ and $HCl_{(aq)}$ compare? Calculate the percent difference based on your experimental results, and comment on the possible sources of that difference.

	Equation	ΔH Value
Investigation 2 Reaction		
Investigation 3 Reaction		
Combined Reaction and Predicted ΔH Value		

Extra Credit

If the heat of reaction, ΔH, was experimentally determined for each of the following reactions:

$$HCl_{(aq)} + NaOH_{(aq)} \rightarrow NaCl_{(aq)} + H_2O \qquad\qquad \Delta H = -13.36 \text{ kcal/mole}$$

$$HCl_{(aq)} + Mg(OH)_{2(aq)} \rightarrow MgCl_{2(aq)} + 2\,H_2O \qquad\qquad \Delta H = -6.00 \text{ kcal/mole}$$

what would be the predicted heat of reaction, ΔH, for the reaction:

$$MgCl_{2(aq)} + 2\,NaOH_{(aq)} \rightarrow 2\,NaCl_{(aq)} + Mg(OH)_{2(aq)} \qquad\qquad \Delta H = \text{_____}$$

Use the space below to show your calculations.

Could you really mix dangerous acids and bases and drink the results?

Exploration of Acids and Bases

Background Concepts

Acids and bases play important roles in our daily lives. For example, hydrochloric acid (HCl) plays a major role in the digestive processes that occur in our stomachs. Window cleaners often contain a base in the form of aqueous ammonia (NH_4OH). Oven and drain cleaners typically contain the very caustic base sodium hydroxide (NaOH). Baking powder and baking soda are added to recipes where they react with acids to form CO_2 gas bubbles to make the batter rise. Acetic acid in the form of vinegar ($HC_2H_3O_2$) is frequently recommended by manufacturers for removing hard water deposits (lime) from the inside of coffee makers, teakettles, and steam irons. Products sold to remove hard water deposits from bathroom fixtures often contain phosphoric acid, H_3PO_4. Now that we know some of the useful ways in which acids and bases can serve, we might ask the less obvious question, "What are they?"

Long ago humans recognized that some substances have a sharp smell and sour taste. If you have ever smelled and tasted vinegar, then you have made the same observations. Since the Latin word for sour is *acidus*, substances exhibiting these characteristics were called acids. Hundreds of years later laboratory experiments repeatedly demonstrated that when certain metals were placed in acidic solutions, hydrogen gas was always one of the products. This association between acid behavior and hydrogen gas eventually led Svante August Arrhenius in 1884 to propose the first successful definition for acids. According to the definition, acids are substances that are capable of releasing hydrogen ions (H^+) in water. For example, when hydrogen chloride gas is bubbled into water the following reaction occurs:

$$HCl_{(g)} + H_2O \rightarrow H_3O^+ + Cl^-$$

Although hydronium ion (H_3O^+) is what actually forms, it is common to eliminate the water from both sides of the equation resulting in the simpler equation:

$$HCl \rightarrow H^+ + Cl^-$$

Since the ionization of HCl in water results in the formation of a hydrogen ion (H^+), then according to the Arrhenius definition, HCl is an acid. Other substances that exhibit similar behavior are:

Acetic acid (vinegar)	$HC_2H_3O_2 \rightarrow H^+ + C_2H_3O_2^-$
Nitric acid	$HNO_3 \rightarrow H^+ + NO_3^-$
Sulfuric acid (battery acid)	$H_2SO_4 \rightarrow 2\,H^+ + SO_4^{2-}$
Phosphoric acid	$H_3PO_4 \rightarrow 3\,H^+ + PO_4^{3-}$

It should be noted that some substances are capable of producing more than one hydrogen ion per molecule. These acids are referred to as diprotic and triprotic acids. Although the name may seem strange, remember that a hydrogen ion (H^+) is a hydrogen atom minus its one and only electron. That leaves only the nucleus, which in most hydrogen atoms consist of a single proton. Hence, most hydrogen ions are just a proton and the terms diprotic and triprotic are just indicating whether they produce two or three separate hydrogen ions when they ionize.

Years of experimentation have also revealed substances that can neutralize the characteristics of acids. These substances are called bases. Arrhenius defined a base as any substance that releases one or more hydroxide ions (OH^-) in water. For example, when sodium hydroxide (NaOH) is dissolved into water, the following ions are produced:

$$NaOH \rightarrow Na^+ + OH^-$$

Note that the water has again been eliminated from the equation for purposes of simplification. As shown below, some bases can produce more than one hydroxide ion. In a fashion similar to acids, these bases can be referred to as dibasic or tribasic.

Potassium hydroxide	KOH	$\rightarrow K^+ + OH^-$
Ammonium hydroxide (aqueous ammonia)	NH_4OH	$\rightarrow NH_4^+ + OH^-$
Magnesium hydroxide (milk of magnesia)	$Mg(OH)_2$	$\rightarrow Mg^{2+} + 2\,OH^-$
Aluminum hydroxide	$Al(OH)_3$	$\rightarrow Al^{3+} + 3\,OH^-$

In more recent years, several other useful acid/base definitions have been proposed; e.g., Brønsted–Lowry and Lewis. However, the Arrhenius definition remains the simplest and is the most commonly used for discussing aqueous (water) solutions.

It is important to understand that regardless of the definition used, acidic and basic solutions always exhibit certain physical and chemical characteristics. Some of these characteristics may be the same. For example, both acid and basic solutions contain ions; therefore, both will conduct electricity. Solutions that conduct electricity are known as electrolytes, a term commonly applied to the sulfuric acid solution in a car's battery. All acidic and basic solutions are, therefore, electrolytes. The degree to which a molecule releases ions in water varies. Some acid and base molecules undergo nearly complete (~100%) ionization forming solutions that contain many ions. Other acid and base molecules experience only a small degree of ionization (<1%) when placed in water. Substances that ionize nearly completely are referred to as strong acids or strong bases and those that have low percentages of ionization are considered weak acids or weak bases. It is important to understand that in this context the terms "strong" and "weak" apply only to the percent of ionization. They should never be confused with the term concentration. Regardless of their concentration and degree of ionization, acid and basic solutions each exhibit characteristic behaviors. The only differences are in the speed of the reaction which, in turn, depends only on the number of H^+ or OH^- ion present.

Careful laboratory experimentation has revealed that solutions containing an excess of H^+ ions exhibit the following characteristics. Acids:

— taste sour;

— turn blue litmus red;

— react with active metals such as iron, zinc, aluminum, and magnesium forming a salt solution and releasing hydrogen (H_2) gas;

— react with metal carbonates and bicarbonates, e.g., $CaCO_3$ and $NaHCO_3$, to produce a salt solution, water, and carbon dioxide (CO_2) gas;

— react with metal oxides and hydroxides, e.g., Na_2O and KOH, to form salt solutions and water.

On the other hand, solutions containing an excess of OH^- ions exhibit the following characteristics. Bases:

— taste bitter;

— feel slippery to the touch;

— turn red litmus paper to blue;

— react with acids to form salt solutions and water.

This last characteristic of bases is referred to as neutralization. Most people know that acids and bases neutralize one another, but they may not know why. Let's look at a simple explanation. In aqueous solutions, $HCl \rightarrow H^+ + Cl^-$ and $NaOH \rightarrow Na^+ + OH^-$, when hydrochloric acid and sodium hydroxide solutions are mixed, the following reaction occurs:

$$H^+ + Cl^- + Na^+ + OH^- \rightarrow Na^+ + Cl^- + H_2O + energy$$

Elimination of the ions that were not changed by the process, called spectator ions, leaves

$$H^+ + OH^- \rightarrow H_2O + energy$$

From these equations, it can be seen that when equal numbers of H^+ and OH^- ions are present, water and energy are the products. (More will be said about the energy produced later in the discussion.) Solutions that contain a balance of H^+ and OH^- ions are called neutral. It may now be obvious that the balance between H^+ and OH^- ions is what determines if a solution is acidic, basic, or neutral. Recognizing the importance of this difference, the Danish biochemist Sørenson, in 1907, developed a mathematical formula that would convert small H^+ ion concentrations into a more convenient number. The numbers are called pH. The use of pH is, therefore, just another way of expressing the concentration of hydrogen ions in a solution.

Before introducing the mathematical formula for the calculation of pH, several other things need a brief explanation. First, there are always some hydrogen and hydroxide ions, even in pure water. Due to their random and chaotic motion, some water molecules experience collisions that result in them being broken into one hydrogen ion and one hydroxide ion. Once formed, these ions eventually bump into oppositely charged ions forming new water molecules. This breaking and reforming of water molecules results in an equilibrium between the water molecules, hydrogen, and hydroxide ions. The equation for this equilibrium is:

$$H_2O \leftrightarrow H^+ + OH^-$$

Examination of the equation reveals that although hydrogen and hydroxide ions are formed, their numbers are equal and the solution therefore remains neutral. Second, experiments measuring the electrical conductivity of water reveal that the actual number of these ions is very small. The concentrations of the hydrogen and hydroxide ions at 25°C are 0.0000001 M or 1×10^{-7} M each. Stated differently, the concentrations are $[H^+] = 1 \times 10^{-7}$ and $[OH^-] = 1 \times 10^{-7}$, where the [ ] indicate that the concentrations are expressed in moles/liter (M).

Finally, substituting the ion concentrations into the equilibrium constant expression for water (K_w) produces the following:

$$K_w = [H^+][OH^-]$$

$$K_w = [1 \times 10^{-7}][1 \times 10^{-7}]$$

$$K_w = 1 \times 10^{-14}$$

From this equation, it is apparent that the hydrogen and hydroxide ion concentrations are inversely proportional. That means that if the hydrogen concentration is increased by adding acid, then the hydroxide ion concentration must be proportionally decreased, so K_w remains constant.

The mathematical relationship Sørenson found that would change the hydrogen concentration into a convenient number is defined as the negative of the logarithm of the hydrogen ion concentration. Mathematically, the formula can be expressed as the following:

$$pH = -\log[H^+] \quad \text{or} \quad [H^+] = 1.0 \times 10^{-pH}$$

Use of a hand-held calculator will quickly convince you that if you enter the number 1.0×10^{-7} and then push the Log key, your display will show a −7.00. By using the +/− key next, the minus sign in the formula will eliminate the negative and the pH becomes 7.00. In this example, since the value of the $[H^+]$ is an even multiple of ten, the second formula could have been used, yielding $[H^+] = 1.0 \times 10^{-7}$. Therefore the pH = 7.00.

As another example, solve the following problem using your calculator.

$$[H^+] = 2.3 \times 10^{-4}$$

$$pH = -\log[H^+]$$

$$pH = -\log(2.3 \times 10^{-4})$$

$$pH = 3.64$$

As can be seen from the above example, when the $[H^+] > [OH^-]$, then the resulting pH values will be less than 7.00. In a similar manner, if $[H^+] < [OH^-]$, then the pH values will be greater than 7.00.

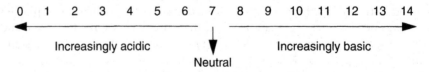

Now that we understand what pH is measuring, the next question is how it is determined. There are two methods in common practice: the pH meter and the use of organic dyes called indicators. The pH meter relies on the small changes in electrical conductivity that take place within a semi-permeable glass electrode when it is placed in solutions of varying numbers of hydrogen ions.

For years, indicators have been used to quickly determine the pH of solutions. You may have already used one of these to determine the pH of your swimming pool water. The color of an individual indicator can only tell you if the pH is above or below specific points. When several indicators are used in conjunction, however, the combined colors can be compared to a color standard that allows for the successful determination of the solution's pH. For example, by using a mixture of the indicators contained in the table below, if the pH is very low (pH = 2) the individual indicators would be red, yellow, or colorless. When these are added, the resulting mixture of indicators would be a red-orange color. On the other hand, if the pH is very high (pH = 10) the individual indicators would be violet, yellow, blue, and red. A combination of these colors would result in a blue-violet color. In a similar fashion, universal indicators have been devised that can give distinct colors for each number on the pH scale.

Indicator	Band 1		Band 2		Band 3
Methyl Violet	Yellow	xxxxxxxx		Violet	
Methyl Orange		Red	xxxxx	Yellow	
Methyl Red		Red	xxxxxxxx	Yellow	
Litmus		Red	xxxxxxxxxx	Blue	
Bromthymol Blue		Yellow	xxxxxx	Blue	
Phenol-phthalein		Colorless	xxxxxx	Red	
Alizarin Yellow R		Yellow		xxx	Red

pH scale: 0 1 2 3 4 5 6 7 8 9 10 11 12 13 14

pH

xxx = Transition

In addition to commercially prepared indicators, many naturally occurring substances also exhibit color changes when exposed to varying pHs. The colored pigments in hot tea, cranberry juice, red cabbage leaves, red onions, hydrangea blossoms, and the leaves of the variegated coleus are just a few.

As a final consideration, remember that the neutralization reaction is exothermic. The source of the energy can be explained by the difference in enthalpies of the individual hydrogen and hydroxide ions as compared to a water molecule. The resulting excess energy produces enough heat to measurably change the temperature of the solution. For a strong acid and strong base, that energy has been determined to be 13.4 kcal/mole.

$$H^+ + OH^- \rightarrow H_2O + 13.4 \text{ kcal/mole}$$

As you complete these experiments, you will be able to describe some of the common characteristics of acidic and basic solutions as well as the usefulness of acid-base indicators. You will also be able to measure and calculate the temperature change and heat of neutralization for strong acids and bases. Finally, you may be able to answer that question, "Could you really mix dangerous acids and bases and drink the results?"

Investigation 1

Objective

To explore the effect of pH on the color of red cabbage.

Safety requirements

■ Always wear splash proof safety goggles.

■ Do not touch any chemicals.

■ Wash skin that is exposed to chemicals in cold water. Then notify the teacher.

■ Discard the waste chemicals as directed by your instructor.

Procedure

1. Obtain one red cabbage leaf from the supply area.

2. Remove a small piece (5 cm × 5 cm) of the leaf and tear the piece to expose a long edge. Place 1 drop of 0.1 M HCl along the torn edge. Record the color along the edge after adding the acid.

3. Along the other torn edge add 1 drop of 0.1 M NaOH solution. Record the color along the edge after adding the base.

4. Tear or cut the remaining cabbage leaf into small (2 cm × 2 cm) pieces and place them in a 250 mL beaker. Add about 100 mL of DI water to the beaker and cover with a watch glass. Using a burner or hot plate, boil the solution for about 5 minutes and cool. Using the watch glass to help separate the leaf pieces, pour off and save the resulting red liquid in a separate beaker.

5. Place 20 drops of 0.1 M HCl into the first compartment of your well tray and 20 drops of 0.1 M NaOH into the second compartment.

6. While viewing against a white background, add 5 drops of the red cabbage extract to the first compartment. Record the color.

_____

7. While viewing against a white background, add 5 drops of the red cabbage extract to the second compartment. Record the color.

_____

8. Using a small stream of water from your wash bottle, rinse the well tray and dry.

Interpretation

1. Were the colors on the edge of the red cabbage leaf and of the red cabbage extract the same after adding the 0.1 M HCl solution? How would you explain this?

2. Were the colors on the edge of the red cabbage leaf and of the red cabbage extract the same after adding the 0.1 M NaOH solution? How would you explain this?

3. If 0.1 M HCl has a $[H^+] = 1.0 \times 10^{-1}$, what is the pH of this solution? Show your calculations.

4. If the 0.1 M NaOH solution has a $[OH^-] = 1.0 \times 10^{-1}$, what is the pH of this solution? Remember that this is a $[OH^-]$ concentration and K_w must be used to calculate the resulting $[H^+]$ before the pH can be determined. Show your calculations.

5. What effect would the addition of a teaspoon of vinegar or lemon juice have on a bowl of cooked red cabbage? Explain your answer.

Investigation 2

Objective

To develop a method that can be used to identify unknown acidic and basic solutions.

Procedure

1. Set your well plate on a piece of white paper. Position it so it has 6 columns across and 4 rows down, as shown in Figure 14-1.

2. Using the dropper provided, place 20 drops of 0.1 M HCl in each of the four wells in the first column. (Remember: rows = left to right and columns = top to bottom)

3. Using the dropper provided, place 20 drops of 0.1 M H_2SO_4 in the wells of the second column.

Figure 14–1: Well plate.

4. Using the dropper provided, place 20 drops of 0.1 M $HC_2H_3O_2$ in the wells in the third column.

5. Using the dropper provided, place 20 drops of 0.1 M H_3PO_4 in the wells in the fourth column.

6. Using the dropper provided, place 20 drops of lemon juice in the wells in the fifth column.

7. Using a dropping pipette, place 20 drops of DI or distilled water in the wells in the sixth column.

8. Using a dropping pipette, now add 5 drops of the prepared red cabbage extract to each well in the first horizontal row. Record in Table 14-1A any observations, reactions, or color changes that occur.

9. Using a spatula, add a small amount (about the size of a match head) of sodium bicarbonate ($NaHCO_3$) to each well in the second horizontal row. Observe carefully and record any reactions or color changes that occur.

10. Using forceps, add one piece (about 5 mm × 5 mm) of magnesium ribbon (Mg) to each well in the third horizontal row. Observe carefully and record any reactions or color changes that occur.

11. Using a clean glass stirring rod, place one drop from the first well in the fourth horizontal row on both a piece of red litmus and on a piece of blue litmus paper. Rinse your stirring rod and dry it before proceeding to the contents of the next well. Record your observations for both the red and blue litmus reactions.

12. Using the dropper provided, add two drops of universal indicator to each well in the fourth horizontal row. By comparing the colors to the color standard provided, record the pH of each solution.

13. If you are uncertain about any of the results, a test can be repeated. Make sure that all wells have been rinsed in DI or distilled water and dried before using again.

14. Leave the column in Table 14-1A, labeled Unknown, blank at this time. Dispose of the liquids in the well tray as recommended by your instructor and rinse with DI or distilled water. Dry the well tray before proceeding with the next part of this Investigation.

	Column 1 0.1 M HCl	Column 2 0.1 M H_2SO_4	Column 3 0.1 M $HC_2H_3O_2$	Column 4 0.1 M H_3PO_4	Column 5 Lemon Juice	Column 6 H_2O	Unknown _____
Row 1 Cabbage							
Row 2 $NaHCO_3$							
Row 3 Mg							
Row 4 Red Litmus							
Row 4 Blue Litmus							
Row 4 Indicator pH							

Table 14–1A: Investigation 2 data.

15. Using the same procedure as in steps 1-7 above, place 20 drops of the following basic solutions in each well in the column.

   — Column 1   0.1 M NaOH

   — Column 2   0.1 M KOH

   — Column 3   0.1 M Ammonia water (NH_4OH)

   — Column 4   0.1 M $NaHCO_3$

   — Column 5   Bleach solution

   — Column 6   DI or distilled H_2O

16. Using the same substances as in steps 8 – 13 above, completeTable 14-1B.

	Column 1 0.1 M NaOH	Column 2 0.1 M KOH	Column 3 0.1 M NH_4OH	Column 4 0.1 M $NaHCO_3$	Column 5 Bleach Solution	Column 6 H_2O	Unknown _____
Row 1 Cabbage							
Row 2 $NaHCO_3$							
Row 3 Mg							
Row 4 Red Litmus							
Row 4 Blue Litmus							
Row 4 Indicator pH							

Table 14–1B: Investigation 2 data.

Interpretation

1. After reviewing your findings in Table 14-1A, what reactions/color changes appear to be characteristic for all acidic solutions?

2. After reviewing your findings in Table 14-1B, what reactions/color changes appear characteristic for all basic solutions?

3. List any substance(s) that did not fit into either the acidic or basic solutions category. Explain your findings.

4. List those characteristics that would enable you to distinguish between an acidic and basic solution.

5. Select one of the unknown solutions provided. Be sure to record in Tables 14-1A and 14-1B the one you used. Using a clean, dry well plate, perform the necessary tests to complete the Unknown column on both Tables 14-1A and 14-1B. Determine if the unknown is an acidic or basic solution. Based on the properties you identified in question 3, explain your conclusion.

Investigation 3

Objective

To determine the amount of energy released by the neutralization process.

Procedure

1. Obtain between 9.5 – 10 mL of a 1.0 M HCl solution and place in a 10 mL graduated cylinder.

2. Using a thermometer, determine the temperature of the HCl solution to the nearest ± 0.1°C and record it in Table 14-2 below.

3. Remove the thermometer, determine the volume of the solution remaining to the nearest ± 0.1 mL and record it in Table 14-2.

4. Transfer the HCl to a clean dry styrofoam cup and place the cup and its contents in a 250 mL beaker.

5. Rinse the 10 mL graduated cylinder with DI or distilled water, dry, and obtain between 9.5 – 10 mL of the 1.0 M NaOH solution. (Attempt to have very nearly the same volumes of acid and base.)

6. Using a clean dry thermometer, determine the temperature of the base solution to the nearest ± 0.1°C and record it in Table 14-2.

7. Remove the thermometer, determine the volume of the solution remaining to the nearest ± 0.1 mL and record it in Table 14-2.

8. Transfer the HCl solution into the styrofoam cup. At your partner's signal, transfer the NaOH solution into the cup containing the HCl solution. Have the partner note the time of transfer and record this as 0 seconds. Use the average temperature of the two solutions mixed for 0 seconds.

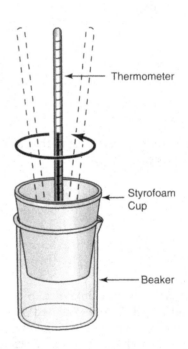

Figure 14–2: Styrofoam cup, beaker, and thermometer.

9. As shown in Figure 14-2, use the thermometer to slowly stir the contents of the cup. Have your partner call out the time every 15-seconds and record the temperatures you report. Continue this process for 3 minutes. Record all temperatures in Table 14-2.

10. Determine the pH of the mixture in the styrofoam cup by transferring a small amount of it into a clean, dry well and adding 2 drops of the universal indicator solution. Record the pH in Table 14-2.

Solution	Temperature, °C		Volume, mL
1.0 M HCl			
1.0 M NaOH			
HCl/NaOH Mixture	**Time, sec**		**Temperature, °C**
	0		(average)
	15		
	30		
	45		
	60		
	75		
	90		
	105		
	120		
	135		
	150		
	165		
	180		
Final pH of HCl/NaOH Mixture			

Table 14–2: Investigation 3 data.

Interpretation

1. Did the temperature of the two solutions increase or decrease when mixed? Using your own words, explain why.

2. Prepare Graph 14-1 using the acid/base mixing data.

3. What is the average temperature of the acid and base solutions, before mixing?

_____ °C

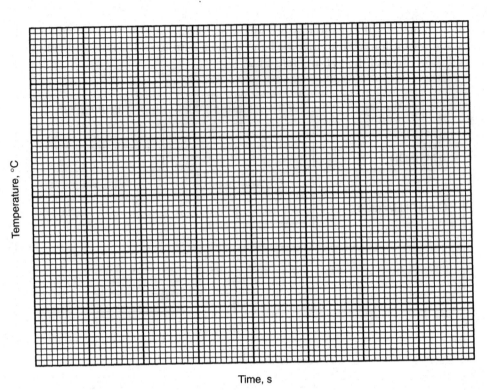

Graph 14–1: Results of mixing the HCl and NaOH solutions.

4. What is the total volume of the acid and base solutions mixed?

_____ mL

5. Use Graph 14-1 to determine the highest temperature reached by the mixture.

_____°C

6. Using the difference between the average temperature before mixing and the highest temperature reached, calculate the temperature change, Δt.

_____°C

7. Assuming that 1.0 mL of acid or base solution weighs 1.0 grams and that the specific heat (sp ht) of water is 1.0 cal/g · °C, calculate the total number of calories produced by this neutralization. Show your work.

$$\text{heat} = \text{sp ht} \times g_{H_2O} \times \Delta t \qquad\qquad = \text{_____ calories}$$

8. Consider the mixture's final pH and explain.

Report Sheet

Experiment 14

Name: _____

Date: _____ Section: _____

1. Using the information collected in this experiment, explain the differences between the characteristics of acidic and basic solutions.

   a. Which unknown did you use in Investigation 2, Interpretation 5. _____

   b. Was the unknown an acid or a base? _____

2. If you were home and needed to determine if a solution was acidic or basic, how would you do it?

3. Suppose you had a universal indicator solution composed of those indicators listed in the table. Demonstrate your understanding of indicators by predicting the colors for solutions with a pH = 3, pH = 7.5, and pH = 10.5. Explain your answers.

4. Using the data collected during neutralization, provide the following:

   — The average temperature of the acid and base solutions, before mixing? _____ °C

   — The total volume of the acid and base solutions? _____ mL

   — The highest temperature reached by the mixture. _____ °C

   — The calculated Δt for the reaction. _____ °C

   — In the space below show how you calculated the number of calories produced. Be sure to include all units in your calculations.

5. Although a basic rule in the chemistry laboratory is to never touch or taste chemicals, attempt to answer the question posed at the beginning of this experiment, "Could you really mix dangerous acids and bases and drink the results?" Explain your answer based on the findings of this experiment and include why it would be a good idea to wait several minutes after mixing to drink the results.

Extra Credit

Based on your measurements, calculate the number of kilocalories of energy that would have been produced if 1 mole of water is produced by a neutralization reaction. Use the number of calories calculated in question 7 of Investigation 3, and the volume and molarity of the solutions to calculate your answer. Use the information presented in Investigation 1 to calculate the percentage error in your calculated results. Discuss possible error sources.

Experiment

15

Is there a difference between a strong and a concentrated acid?

Acid Concentrations and Strengths

Background Concepts

Acid and base chemistry is fundamental to understanding many industrial processes, environmental balances, and biological functions. Arrhenius defined an acid as a substance that donates hydrogen ions, H^+ in water. The companion definition for a base is a substance that donates hydroxide ions, OH^- in water.

The reaction of an acid with a base is called neutralization. The general formula for the reaction is:

$$H^+Nm^- + M^+OH^- \longrightarrow M^+Nm^- + H{-}O{-}H$$

where M^+ is a metal ion and Nm^- is a nonmetal or polyatomic ion. The products of a neutralization reaction are always a salt (M^+Nm^-) and water (H_2O).

In pure water at room temperature water molecules suffer constant collisions. Sometimes these collisions result in a water molecule being ionized (broken) into one hydrogen ion and one hydroxide ion. In the next instant, these ions may again unite and reform a water molecule. An equilibrium exists between the number of hydrogen ions, hydroxide ions, and water molecules. At room temperature, the number of hydrogen and hydroxide ions present are $[H^+] = 1.0 \times 10^{-7}$ and $[OH^-] = 1.0 \times 10^{-7}$. The equilibrium constant for water, K_w, is given by the expression $K_w = [H^+][OH^-]$ and $K_w = 1.0 \times 10^{-14}$. The important thing to understand about this relationship is that if the $[H^+]$ of a solution increases because an acid is added, then the $[OH^-]$ must become proportionally smaller, and vice versa.

The concentration of hydrogen ions in a solution is what is being measured by the pH scale, which ranges from 1 to 14. It is mathematically defined as $pH = -\log [H^+]$. This means that pH = 1 is very acidic and has a very large hydrogen ion concentration, while pH = 14 is very basic and has a very low hydrogen ion concentration. Pure water has a pH = 7, meaning that it is neutral and has an exact balance of hydrogen and hydroxide ions.

If the ratio of hydrogen and hydroxide ions in a solution determines its pH, then it would seem logical that adding one mole of any acid would have the same effect on the pH of the solution. This, however, is not the case, since there are both *strong* and *weak* acids, as well as *monoprotic* and *diprotic* acids. By strong acids, we mean that they ionize 100% and yield as many hydrogen ions as there were acid molecules. Hydrochloric acid is an example of such a strong acid, and ionizes 100% in dilute solutions. Therefore, a 0.010 M solution of HCl would have both a H^+ and Cl^- ion concentration of 0.010 M, and a pH = 2.

$$0.010 \text{ M} \times 100\% \quad = \quad 0.010 \text{ M} \quad \quad 0.010 \text{ M}$$

$$HCl \quad \longrightarrow \quad H^+ \quad + \quad Cl^-$$

$$pH = -\log[H^+]$$

$$pH = -\log[0.010]$$

$$pH = -(-2.0)$$

$$pH = 2.0$$

When a weak acid is used, however, only a part of the molecules ionize, so that the number of hydrogen ions in the solution is much less than the number of acid molecules added. Acetic acid (vinegar, $HC_2H_3O_2$) is such a weak acid, and ionizes to only about 4%. In a 0.010 M solution of $HC_2H_3O_2$, both the H^+ and $C_2H_3O_2^-$ ion concentrations would be 4% (0.010 M) or 0.0004 M and it would have a pH = 3.3.

$$0.010 \text{ M} \times 4\% \quad = \quad 0.0004 \text{ M} \quad \quad 0.0004 \text{ M}$$

$$HC_2H_3O_2 \quad \longrightarrow \quad H^+ \quad + \quad C_2H_3O_2^-$$

$$pH = -\log[H+]$$

$$pH = -\log[0.0004]$$

$$pH = -(-3.3)$$

$$pH = 3.3$$

There are also *strong* and *weak* bases. In a strong base, 100% of the molecules are ionized and in a weak base only about 1%. Sodium hydroxide (Drano, NaOH) is a strong base and ionizes 100% in dilute solutions, while ammonium hydroxide (ammonia water, NH_4OH) is a weak base. One way to determine whether acids and bases are strong or weak is by comparing the pH of equal molar solutions. It should also be noted, however, that *weak* acids and bases will still neutralize equal volumes of equal molar *strong* acids and bases.

The second factor that determines the pH of an acid or base solution depends on the number of H^+ or OH^- ions released per molecule. Monoprotic or monobasic molecules release only one H^+ or OH^- per molecule. As we can see from the above examples, both HCl and $HC_2H_3O_2$, produce only one hydrogen ion when a molecule ionizes. Diprotic or dibasic acids and bases release two H^+ or OH^- per molecule. Therefore the concentration of the H^+ or OH^- ions may be as much as twice that of the molecule concentration, depending on whether it is a strong or weak acid or base. Sulfuric acid, for example, is a strong diprotic acid. The H^+ concentration in a solution of 0.010M H_2SO_4 would be 0.02 M and have a pH = 1.7.

$$0.010 \text{ M} \times 100\% = 2\,(0.010 \text{ M}) \qquad 0.010 \text{ M}$$

$$H_2SO_4 \longrightarrow 2\,H^+ \quad + \quad SO_4{}^{2-}$$

$$pH = -\log[H^+]$$

$$pH = -\log[0.02]$$

$$pH = -(-1.698)$$

$$pH = 1.7$$

From this, it can be seen that a 0.010 M solution of the strong diprotic sulfuric acid would have a $[H^+]$ that is twice the original molecular concentration. A similar example could be used for the strong dibasic $Ca(OH)_2$, in which the OH^- concentration would be twice that of the original $Ca(OH)_2$ concentration.

Finally, an indicator is an organic substance whose color is pH dependent. Bogen's Universal Indicator is a mixture of indicators that were selected in such a way that they produce a color variation for each pH. A color chart will be provided in the laboratory. However, in general, the colors corresponding to the pHs are:

pH	1	2	3	4	5	6	7	8	9	10	11	12	13	14
Color	Red	Red-Orange	Orange	Orange-Yellow	Yellow	Yellow-green	Lime-green	Green	Green-Blue	Blue	Blue-violet	Violet	Violet-purple	Purple

In this experiment, we will use a titration technique and Bogen's Universal Indicator solution to investigate the strength and number of ions released per molecule of several different acids.

Investigation 1

Objective

To determine the concentration and volume ratios required to neutralize a strong monoprotic acid with a strong monobasic base.

> ### Safety Requirements
>
> ■ Always wear splash proof safety goggles.
>
> ■ Avoid getting acid and base solution on your skin, your clothes, and especially in your eyes.
>
> ■ Return all waste solutions to the designated waste container, as directed by your instructor.
>
> ■ Never touch chemicals or put anything in your mouth while in the laboratory.

Procedure

> **Note:** It is expected that this experiment will be performed in pairs. One person may initially perform the experiment, while the other records the data. These roles should frequently be reversed so both participants can gain equal experience in performing the experiment. Each participant should obtain a complete set of the data collected. The fact that you have recorded the same data does not mean that you must arrive at the same conclusions.

1. Obtain a clean, dry 24-well plate and a supply of plastic droppers so that you will have one for each of the acid and base solutions as well as one for the Bogen's Universal Indicator solution. As directed by your instructor, you will also need a supply of glass rods, plastic coffee stirrers, or toothpicks to stir the well contents.

2. Obtain a supply of 0.010 M NaOH and a bottle of indicator for your work area.

3. Obtain a supply of the four hydrochloric acid solutions (0.010 M HCl, 0.020 M HCl, 0.050 M HCl, and 0.100 M HCl).

4. Place your clean 24-well plate on a piece of white paper.

5. Using a plastic dropper, add 5 drops of the 0.010 M HCl solution to the well in the upper left of the 24-well plate, as shown in Figure 15-1.

Figure 15-1: Well plate.

6. In a similar fashion, add 5 drops of each of the other HCl solutions to the remaining wells along the top row.

7. Add 1 drop of Bogen's Universal Indicator solution to each of the wells. Note the color of the solution in each well, estimate and record the pH of each solution in Table 15-1.

8. Titrate the contents of the first well (0.010 M HCl) by carefully adding 0.010 M NaOH solution by the drop. Stir the mixture after each drop is added. Count the number of drops it takes to reach or pass by the lime-green color. (It is permissible to estimate to the nearest $\frac{1}{2}$ drop.)

9. Record in Table 15-1 the number of drops of 0.010 M NaOH solution required to neutralize (pH = 7) the 5 drops of the 0.010 M HCl solution.

10. Repeat steps 8 and 9 for each of the other HCl solutions.

	0.010 M HCl	0.020 M HCl	0.050 M HCl	0.100 M HCl
pH of HCl solution				
Drops of HCl used				
Drops of 0.010 M NaOH				

Table 15–1: HCl titration results.

Interpretation

1. Examine the data collected for the titration of each HCl solution. Does the number of drops of NaOH required to reach pH = 7 increase for each of the wells?

2. Is there a correlation between the number of drops of NaOH required and the concentration of the HCl? In other words, if 0.020 M HCl is twice as concentrated as 0.010M HCl, did it take twice as many drops of NaOH to neutralize it?

3. When you titrate an acid with a base, base is added until the number of hydroxide ions equals the number of hydrogen ions (pH = 7). That means that the concentration of strong monoprotic acid times the drops used must equal the concentration of strong monobasic base times the drops used. Table 15-2 is partially completed. Make your calculations based on this relationship and then transfer your data from Table 15-1 to complete the table.

Concentration of acid (M)	No. of drops of acid used	Concentration of base (M)	Predicted drops of base needed	Actual drops of base used
0.010 M HCl	5 drops	0.010 M NaOH	5 drops	
0.020 M HCl	5 drops	0.010 M NaOH		
0.050 M HCl	5 drops	0.010 M NaOH		
0.100 M HCl	5 drops	0.010 M NaOH		

Table 15–2: Predicted and actual number of drops required for neutralization.

4. How well do the predicted number of drops of base needed and actual drops of base used compare? Attempt to explain any differences you may have found.

Investigation 2

Objective

To determine the concentration and volume ratios required to neutralize a weak monoprotic acid with a strong monobasic base.

Procedure

1. Use the next row of clean dry wells below the neutralized HCl for these tests.

2. Obtain a supply of the four acetic acid solutions (0.010 M $HC_2H_3O_2$, 0.020 M $HC_2H_3O_2$, 0.050 M $HC_2H_3O_2$, and 0.100 M $HC_2H_3O_2$).

3. Using a plastic dropper, add 5 drops of the 0.010 M $HC_2H_3O_2$ solution to the well on the left side.

4. In a similar fashion, add 5 drops of each of the other $HC_2H_3O_2$ solutions to the remaining wells along this row.

5. Add 1 drop of Bogen's Universal Indicator solution to each of the wells. Note the color of the solutions in each well, estimate and record the pH of each solution in Table 15-3.

6. Titrate the contents of the first well (0.010 M $HC_2H_3O_2$) by carefully adding 0.010 M NaOH solution by the drop. Stir the mixture after each drop is added. Count the number of drops used to reach or pass by the lime-green color. (It is permissible to estimate to the nearest $\frac{1}{2}$ drop.)

7. Record in Table 15-3 the number of drops of 0.010 M NaOH solution required to neutralize (pH = 7) the 5 drops of the 0.010 M $HC_2H_3O_2$ solution.

8. Repeat steps 6 and 7 for each of the other $HC_2H_3O_2$ solutions.

	0.010 M $HC_2H_3O_2$	0.020 M $HC_2H_3O_2$	0.050 M $HC_2H_3O_2$	0.100 M $HC_2H_3O_2$
pH of $HC_2H_3O_2$ solution				
Drops of $HC_2H_3O_2$ used				
Drops of 0.010 M NaOH				

Table 15–3: $HC_2H_3O_2$ titration results.

Interpretation

1. Examine the data collected for the titration of each $HC_2H_3O_2$ solution. Do the number of drops of NaOH required to reach pH = 7 increase for each of the wells?

2. Referring to Table 15-1, how do the number of drops required to neutralize equal amounts of equal concentrations $HC_2H_3O_2$ and HCl compare?

3. Table 15-4 is partially completed. Make the necessary calculations and then transfer your data from Table 15-3 to complete the table.

Concentration of acid (M)	No. of drops of acid used	Concentration of base (M)	Predicted drops of base needed	Actual drops of base used
0.010 M $HC_2H_3O_2$	5 drops	0.010 M NaOH	5 drops	
0.020 M $HC_2H_3O_2$	5 drops	0.010 M NaOH		
0.050 M $HC_2H_3O_2$	5 drops	0.010 M NaOH		
0.100 M $HC_2H_3O_2$	5 drops	0.010 M NaOH		

Table 15–4: Predicted and actual number of drops required for neutralization.

4. How well do the predicted number of drops of base needed and actual drops of base used compare? Attempt to explain any differences you may have found.

5. How does the number of drops of base required to neutralize equal volumes and concentrations of monoprotic strong and weak acids compare?

6. Are there any differences between the pH of equal concentrations of monoprotic strong and weak acids? How can you explain this difference?

Investigation 3

Objective

To determine the concentration and volume ratios required to neutralize a diprotic acid with a strong monobasic base.

Procedure

1. Select the next row of clean, dry wells.

2. Obtain a supply of the four solutions of oxalic provided (0.010 M $H_2C_2O_4$, 0.020 M $H_2C_2O_4$, 0.050 M $H_2C_2O_4$, and 0.100 M $H_2C_2O_4$).

3. Using a plastic dropper, add 5 drops of the 0.010 M $H_2C_2O_4$ solution to the first well.

4. In a similar fashion, add 5 drops of each of the other $H_2C_2O_4$ solutions to the remaining wells along the row.

5. Add 1 drop of Bogen's Universal Indicator solution to each of the wells. Note the color of the solutions in each well, estimate and record the pH of the solution in Table 15-5.

6. Titrate the contents of the first well (0.010 M $H_2C_2O_4$) by carefully adding 0.010 M NaOH solution by the drop. Stir the mixture after each drop is added. Count the number of drops to reach or pass by the lime-green color. (It is permissible to estimate to the nearest $\frac{1}{2}$ drop.)

7. Record in Table 15-5 the number of drops of 0.010 M NaOH solution required to neutralize (pH = 7) the 5 drops of the 0.010 M $H_2C_2O_4$ solution.

8. Repeat steps 6 and 7 for each of the other $H_2C_2O_4$ solutions.

	0.010 M $H_2C_2O_4$	0.020 M $H_2C_2O_4$	0.050 M $H_2C_2O_4$	0.100 M $H_2C_2O_4$
pH of $H_2C_2O_4$ solution				
Drops of $H_2C_2O_4$ used				
Drops of 0.010 M NaOH				

Table 15–5: $H_2C_2O_4$ titration results.

Interpretation

1. Examine the data collected for the titration of the various $H_2C_2O_4$ samples. Do the number of drops of NaOH required to reach pH = 7 increase for each of the wells?

2. Referring to Table 15-1, how do the number of drops of NaOH required to neutralize equal amounts of equal concentrations $H_2C_2O_4$ and HCl compare?

3. Using the comparison found in question 2, complete Table 15-6 by predicting the number of drops of NaOH required to neutralize the drops of acid used. Transfer the actual number of drops of NaOH used from Table 15-5.

Concentration of acid (M)	No. of drops of acid used	Concentration of base (M)	Predicted drops of base needed	Actual drops of base used
0.010M $H_2C_2O_4$	5 drops	0.010M NaOH		
0.020 M $H_2C_2O_4$	5 drops	0.010M NaOH		
0.050M $H_2C_2O_4$	5 drops	0.010M NaOH		
0.100 M $H_2C_2O_4$	5 drops	0.010M NaOH		

Table 15–6: Predicted and actual number of drops required for neutralization.

4. How well do the predicted number of drops of base needed and actual drops of base used compare? Attempt to explain any differences you may have found.

5. Compare Tables 15-2 and 15-6. What relationship, if any, do you find between the amount and concentration of NaOH required to neutralize equal concentrations and drops of the two acids?

6. Based on your findings in question 7, would you identify oxalic acid as a monoprotic or a diprotic acid? Explain your answer.

7. Compare Tables 15-1 and 15-5. What relationship, if any, do you find between the concentration of the acid solutions and their pH? Explain your answer.

Name: _____

Date: _____ Section: _____

1. Complete the following table by extracting the necessary information from Tables 15-1, 15-3, and 15-5.

	0.010 M HCl	0.020 M HCl	0.050 M HCl	0.100 M HCl
pH of HCl solution				
Drops of HCl used				
Drops of 0.010 M NaOH				
	0.010 M $HC_2H_3O_2$	0.020 M $HC_2H_3O_2$	0.050 M $HC_2H_3O_2$	0.100 M $HC_2H_3O_2$
pH of $HC_2H_3O_2$ solution				
Drops of $HC_2H_3O_2$ used				
Drops of 0.010 M NaOH				
	0.010 M $H_2C_2O_4$	0.020 M $H_2C_2O_4$	0.050 M $H_2C_2O_4$	0.100 M $H_2C_2O_4$
pH of $H_2C_2O_4$ solution				
Drops of $H_2C_2O_4$ used				
Drops of 0.010 M NaOH				

2. Complete the following table by extracting the necessary information from Tables 15-2, 15-4, and 15-6.

Concentration of acid/base (M)	No. of drops of acid/base used	Concentration of acid/base (M)	Predicted drops acid/base needed	Actual drops acid/base used
0.010 M HCl	5 drops	0.010 M NaOH	5 drops	
0.020 M HCl	5 drops	0.010 M NaOH		
0.050 M HCl	5 drops	0.010 M NaOH		
0.100 M HCl	5 drops	0.010 M NaOH		
0.010 M $HC_2H_3O_2$	5 drops	0.010 M NaOH	5 drops	
0.020 M $HC_2H_3O_2$	5 drops	0.010 M NaOH		
0.050 M $HC_2H_3O_2$	5 drops	0.010 M NaOH		
0.100 M $HC_2H_3O_2$	5 drops	0.010 M NaOH		
0.010M $H_2C_2O_4$	5 drops	0.010M NaOH		
0.020 M $H_2C_2O_4$	5 drops	0.010M NaOH		
0.050M $H_2C_2O_4$	5 drops	0.010M NaOH		
0.100 M $H_2C_2O_4$	5 drops	0.010M NaOH		

3. Review the difference in the number of drops of NaOH it took to neutralize equal drops of equal concentration strong and weak monoprotic acids. What difference, if any, did you find?

4. Review the pH of equal concentration strong and weak monoprotic acids. What difference, if any, did you find?

5. Review the difference in the number of drops of NaOH it took to neutralize equal drops of equal concentration monoprotic and diprotic acids. What differences, if any, did you find?

6. Review the pH of equal concentration HCl, $HC_2H_3O_2$, and $H_2C_2O_4$. What differences, if any, did your find?

7. What is the difference between a strong and a concentrated acid?

Extra Credit

1. Calculate the volume of 2.0 M KOH that would be required to neutralize 25.00 mL of 1.0 M HCl.

2. Calculate the volume of 0.10 M H_2SO_4 that would be required to neutralize 50.00 mL of 0.50 M NaOH.

3. If a 0.05 M $HC_2H_3O_4$ solution is 1% ionized, calculate the pH of the solution. How does your answer compare to your experimental results?

What percent of vinegar is acetic acid?

Percent of Acetic Acid in Vinegar: An Acid/Base Titration

Background Concepts

The amount of reactants and products involved in chemical reactions are commonly investigated in two ways: gravimetrically (by mass) and volumetrically (by volume and concentration). In this experiment, we will use both mass and a volumetric method to determine the percent of acetic acid, $HC_2H_3O_2$, in vinegar.

Titration is the name given to the process used to determine the volume of a solution of known concentration needed to react with a given mass or volume of a second substance. From its description, it is apparent that to perform a titration requires a method to accurately measure the volume of the solution(s) used. Figure 16-1 shows some examples of the unique and often expensive glassware that is available to measure liquid volumes. Items such as burets and Mohr pipets

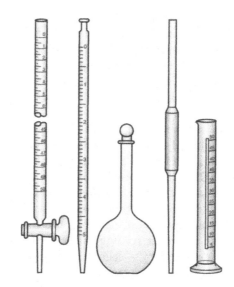

Figure 16-1: Volumetric glassware: (a) buret, (b) Mohr pipet, (c) volumetric flask, (d) pipet, and (e) graduated cylinder.

have the advantage of measuring the volume of liquid delivered over a range of volumes with high accuracy. Other items, such as pipets and volumetric flasks are limited to measuring only a specific volume, such as ten or fifty milliliters. Graduated cylinders are much less expensive and can again measure volumes over a range, but their accuracies are limited by their graduation markings.

Plastic droppers are inexpensive and intended for the transfer of small amounts of liquid. Through a process called calibration, however, the volume of each of its drops can quite accurately be determined. Once the volume of each drop is known and by counting the number of drops delivered, plastic droppers can be used to measure the amount of liquid transferred.

The second necessity for conducting a titration is to have a solution of known concentration. Such solutions are known as standard solutions and can be made using two different methods. In one method, a known amount of a pure substance is weighed out and dissolved in a known amount of solution. In the second method, the solution is first prepared to be near the desired concentration, and then standardized by reacting a known volume of it with a carefully weighed amount of some other substance. Once the concentration of the standard solution is known, it can be used to prepare other standard solutions, or directly used to determine the amount of another substance.

The last necessity for conducting a titration is a means for determining when the amounts of the two reacting substances become stoichiometrically equal. On some occasions, this can be accomplished electronically, based on the change in the electrical conductivity of the solution as it nears a point of chemical equivalency. For example, since acidic solutions contain an excess of hydrogen ions, H^+, the addition of a basic solution containing hydroxide ions, OH^-, will greatly reduce the number of free ions H^+. This, in turn, raises the pH of the solution, eventually reaching a pH = 7 at the end-point. The net ionic equation for the reaction shows that one mole of hydrogen ions is reacted with one mole of hydroxide ions according to the following equation:

$$H^+ + OH^- \longrightarrow H_2O$$

In many applications, a pH meter is now the choice for determining the end-point of acid-base reactions. In addition, a number of indicator dyes, such as methyl orange, bromthymol blue, litmus, and phenolphthalein can be used to detect the end-point by their color changes. For a variety of reasons, phenolphthalein is a popular indicator for many acid and base titrations. It is colorless in acid solution, becomes tinged with pink as the stoichiometric end-point nears, and becomes shocking pink in a base solution.

In this experiment, we will first calibrate a dropper to accurately measure the volume of liquid used. Second, we will use a weighed amount of oxalic acid dihydrate, $H_2C_2O_4 \cdot 2\, H_2O$, to produce a standard solution, which will be used to standardize the NaOH solution provided. Finally, we will use the standardized NaOH solution to determine the weight/volume percent of acetic acid that is present in vinegar.

Titration remains the workhorse of industrial quality control. It is often completely automated, relying on an electronic end-point detection rather than an indicator, and automatic reading and refilling burettes rather than counting the number of drops delivered by a calibrated dropper. Despite these differences, the underlying principles remain the same.

Investigation 1

Objective

To calibrate the volume of a drop from a plastic dropper.

Procedure

1. Obtain a 10 mL graduated cylinder, a small beaker of DI water and the plastic dropper you are going to use for this experiment.

2. Add between 6 and 7 mL of DI water to the graduated cylinder. Keeping your eye level with the bottom of the meniscus as shown in Figure 16-2, read the initial water volume in the cylinder to the nearest ± 0.02 mL and record the information in Table 16-1. Try holding a piece of paper behind the cylinder to "light" the bottom of the meniscus.

3. Squeeze the dropper bulb and allow it to draw up some DI water. Be certain that there are no air bubbles in the dropper tube.

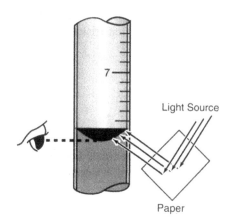

Figure 16-2: With your eye level, read the bottom of the meniscus.

4. While holding the dropper at about a 45° angle, slowly count and add approximately 35 drops of water into the graduated cylinder.

5. Again keeping your eye level with the bottom of the meniscus, read and record in Table 16-1, the final water volume in the cylinder.

6. Empty a part of the water in the graduated cylinder and repeat steps 2 – 5 two more times. You do not have to use the same exact number of drops each time and your starting and ending volumes should not be the same each time. Record your data for each trial in Table 16-1.

	Trial 1	Trial 2	Trial 3
Graduated Cylinder Volume Reading, Final	mL	mL	mL
Graduated Cylinder Volume Reading, Initial	mL	mL	mL
Volume of Water Added	mL	mL	mL
Drops of Water Added	drops	drops	drops
Volume per Drop	mL/drop	mL/drop	mL/drop
Average Volume per Drop			

Table 16–1: Drop calibration data.

Interpretation

1. Complete Table 16-1 by calculating the volume of water added to the cylinder each time.

2. Divide the volume of water added to the cylinder each time by the number of drops used for that trial and record in Table 16-1.

3. Do any of the calculated mL/drop values differ significantly from the other trials? If so, explain, and if necessary omit or repeat that trial.

4. Using the calculated volume per drop for each trial, calculate and record in Table 16-1 the average volume per drop for this dropper.

5. Would you expect the average volume per drop to be the same for all similar plastic droppers using the same liquid? Explain your answer.

Investigation 2

Objective

To make a standard solution by using a known mass of oxalic acid dissolved into a known volume of solution.

Procedure

In this investigation, you are preparing your standard oxalic acid solution, whose accuracy will determine the outcome of the rest of this experiment. Be sure to carefully follow the directions.

1. Obtain a sheet of weighing paper. Fold a sharp crease in the paper from one corner to the other. Place the paper on a centigram balance and either zero the balance with the paper or zero the empty balance and then determine the mass of the paper and record it in Table 16-2.

2. Using a clean or provided spatula, transfer between 0.50 and 0.60 ± 0.01 g of oxalic acid dihydrate, $H_2C_2O_4 \cdot 2\ H_2O$ onto the weighing paper. Read and record the mass of oxalic acid and paper in Table 16-2.

3. Using the fold to help guide the crystals, carefully transfer all of the oxalic acid into your 10 mL graduated cylinder. Add approximately 5 mL of DI water from your wash bottle. Using a thin clean stirring rod, gently stir the contents of your cylinder until all of the oxalic acid crystals have dissolved. This may take several minutes. More DI water may be added to get the crystals dissolve, if necessary.

4. Using a wash bottle, add enough more DI water so that the bottom of the meniscus is near the 10 mL mark. As you add this water, use it to rinse the stirring rod as you remove it so you can determine and record the final volume of just the solution. Once you have read and recorded the final oxalic acid volume, put the stirring rod back into the solution and use it to stir the solution to ensure that it is of uniform concentration from top to bottom.

Mass of weighing paper + oxalic acid	g
Mass of weighing paper	g
Mass of oxalic acid used	g
Final volume of the oxalic acid solution	mL
Molarity of the standard oxalic acid solution	M

Table 16–2: Data for preparing a standard oxalic solution.

Interpretation

1. Determine the mass of oxalic acid dihydrate crystals used and record it in Table 16-2.

2. The molar mass of oxalic acid dihydrate is 126 g/mole, calculate the number of moles of oxalic acid placed into the graduated cylinder.

$$\text{moles} = \frac{\text{grams used}}{\text{molar mass}} = \underline{\hspace{2cm}} = \underline{\hspace{2cm}} \text{ moles } H_2C_2O_4$$

3. What is the final volume of the oxalic acid solution, in liters?

$$\underline{\hspace{2cm}} \text{ mL} \times \frac{1 \text{ liter}}{1,000 \text{ mL}} = \underline{\hspace{2cm}} \text{ liters}$$

4. Using the information from steps 2 and 3, calculate and record in Table 16-2 the molarity, M, of the standard oxalic solution.

$$M = \frac{\text{moles of solute}}{\text{liters of solution}} = \underline{\hspace{2cm}} = \underline{\hspace{1cm}} \frac{\text{moles}}{\text{liters}} = \underline{\hspace{1cm}}$$

Investigation 3

Objective

To standardize an unknown NaOH solution, using the standard oxalic acid solution.

Procedure

1. Place your clean dry well plate over a piece of white paper as shown in Figure 16-3.

2. Rinse your calibrated dropper by removing two small portions of the standard $H_2C_2O_4$ solution and then disposing of the rinse solution.

3. Using your rinsed calibrated dropper, carefully transfer about 35 drops of the standard $H_2C_2O_4$ solution into each of three consecutive wells. Record in Table 16-3 the actual number of drops used in each well. Discard any of the solution remaining in your dropper.

Figure 16-3: Titration equipment.

4. Add one drop of the acid/base indicator, phenolphthalein, to each well. Note its color in an acid solution.

5. Now rinse your calibrated dropper with two portions of DI water and dispose of the rinse water. Then rinse your dropper with two small portions of the NaOH solution you are going to standardize. Discard the rinse solution each time.

6. Proceed to titrate the $H_2C_2O_4$ solution in each. Remove the stirring rod from the graduated cylinder and rinse it with three portions of DI water from your wash bottle. Dry the stirring rod and then use it to stir the contents of the well after each drop of NaOH is added. Carefully count the number of drops of the NaOH solution that is required to produce a pink color and record the number of drops of NaOH solution required to reach the end-point of the titration.

7. Repeat step 6 for each of the remaining wells. Record your data for the titrations in Table 16-3.

	Trial 1	Trial 2	Trial 3
Molarity of $H_2C_2O_4$ solution (Investigation 2)			
Number of drops of $H_2C_2O_4$ solution used			
Volume per drop (Investigation 1)			
Number of mL of $H_2C_2O_4$ solution used			
Number of liters of $H_2C_2O_4$ solution used			
Number of moles of $H_2C_2O_4$ in sample			
Number of drops of NaOH solution used			
Volume per drop (Investigation 1)			
Number of mL of NaOH solution used			
Number of liters of NaOH solution used			
Calculate the number of moles of NaOH present			
Calculated molarity, M, of the NaOH solution			
Average molarity, M, of the NaOH solution			

Table 16–3: Data to standardize the NaOH solution.

Interpretation

1. Using the volume per drop of your calibrated dropper found in Investigation 1, calculate the number of mL and liters of $H_2C_2O_4$ solution used in each of the three trials. Record these volumes in Table 16-3.

2. Using the molarity of the $H_2C_2O_4$ determined in Investigation 2 and the volume, in liters, calculate the number of moles of $H_2C_2O_4$ present in each of the titrations.

3. Using the volume per drop of your calibrated dropper found in Investigation 1, calculate the number of mL and liters of NaOH solution used in each of the titrations. Record these volumes in Table 16-3.

4. According to the equation for the titration reaction $H_2C_2O_4 + 2\ NaOH \longrightarrow Na_2C_2O_4 + 2\ H_2O$, for each mole of $H_2C_2O_4$ used, there must be twice the moles of NaOH present at the end-point. Calculate the number of moles of NaOH present in each trial, using the following conversion:

$$\underline{\hspace{2cm}} \text{ moles } H_2C_2O_4 \times \frac{2 \text{ moles NaOH}}{1 \text{ mole } H_2C_2O_4} = \underline{\hspace{2cm}} \text{ moles NaOH}$$

5. Calculate the molarity, M, of the NaOH solution, based on the results of each titration.

$$\underline{\hspace{1cm}} \frac{\text{moles NaOH}}{\text{liters NaOH}} = \underline{\hspace{1cm}} \frac{\text{moles}}{\text{liters}} = \underline{\hspace{1cm}} M$$

6. If the results of the three titrations are within acceptable agreement, calculate the average molarity of the NaOH solution. Record the average in Table 16-3.

▋ Investigation 4

Objective

To determine the weight/volume percent of acetic acid in vinegar using a standardized NaOH solution.

Procedure

1. Using a clean dry container, obtain a small amount (5 mL) of the vinegar solution.

2. Using the same steps as in Investigation 3, titrate three separate samples of vinegar with the standardized NaOH solution and record your data in Table 16-4. Be sure to rinse the dropper before each change of solution.

	Trial 1	Trial 2	Trial 3
Molarity of NaOH solution (Investigation 2)			
Number of drops of NaOH solution used			
Volume per drop (Investigation 1)			
Number of mL of NaOH solution used			
Number of liters of NaOH solution used			
Number of moles of NaOH in sample			
Number of drops of $HC_2H_3O_2$ solution used			
Volume per drop (Investigation 1)			
Number of mL of $HC_2H_3O_2$ solution used			
Number of liters of $HC_2H_3O_2$ solution used			
Calculate the number of moles of $HC_2H_3O_2$ present			
Calculated molarity, M, of the $HC_2H_3O_2$ solution			
Average molarity, M, of the vinegar ($HC_2H_3O_2$) solution			

Table 16–4: Data to determine the weight/volume percent of acetic acid in vinegar.

Interpretation

1. Using the volume per drop of your calibrated dropper found in Investigation 1, calculate the number of mL and liters of NaOH solution used in each of the three trials. Record these volumes in Table 16-4.

2. Using the molarity of the NaOH determined in Investigation 3, and the volume in liters, calculate the number of moles of NaOH present in each of the titrations.

3. Using the volume per drop of your calibrated dropper found in Investigation 1, calculate the number of mL and liters of $HC_2H_3O_2$ solution used in each of the titrations. Record these volumes in Table 16-4.

4. According to the equation for the titration reaction $HC_2H_3O_2 + NaOH \longrightarrow NaC_2H_3O_2 + H_2O$, for each mole of NaOH used, there is only one mole of $HC_2H_3O_2$ present at the end-point. Calculate the number of moles of $HC_2H_3O_2$ present in each trial, using the following conversion:

$$\underline{\hspace{1.5cm}} \text{ moles NaOH} \times \frac{1 \text{ mole } HC_2H_3O_2}{1 \text{ mole NaOH}} = \underline{\hspace{1.5cm}} \text{ moles } HC_2H_3O_2$$

5. Calculate the molarity, M, of the $HC_2H_3O_2$ solution based on the results of each titration.

$$\frac{\underline{\hspace{1cm}} \text{ moles } HC_2H_3O_2}{\text{liters } HC_2H_3O_2} = \underline{\hspace{1cm}} \frac{\text{moles}}{\text{liters}} = \underline{\hspace{1cm}} \text{ M}$$

6. If the results of the three titrations are within acceptable agreement, calculate the average molarity of the vinegar ($HC_2H_3O_2$) solution. Record the average in Table 16-4.

---

7. Based on the average molarity, M, of the vinegar solution, if acetic acid, $HC_2H_3O_2$, has a molar mass of 60 g/mole, how many grams of acetic acid are present in a liter of vinegar?

$$\underline{\hspace{2cm}} \frac{\text{moles } HC_2H_3O_2}{\text{liter vinegar}} \times \frac{60 \text{ grams } HC_2H_3O_2}{\text{mole } HC_2H_3O_2} = \underline{\hspace{2cm}} \frac{\text{g } HC_2H_3O_2}{\text{liter vinegar}}$$

8. Weight-volume percent is found by dividing the mass of solute by its volume, expressed in milliliters, times 100. Using the following formula, calculate the $\%_{(wt/vol)}$ of acetic acid in vinegar.

$$\frac{\text{g } HC_2H_3O_2}{\text{liter vinegar}} \times 100 = \underline{\hspace{2cm}} \%_{(wt/vol)}$$

Report Sheet

Experiment 16

1. Average volume per drop from the calibrated dropper.

    _____ mL/drop

2. Molarity of the standard oxalic acid, $H_2C_2O_4$, solution.

    _____ M

3. Molarity of the standardized NaOH solution.

    _____ M

4. Transfer the information from Table 16-4 to complete the following table.

	Trial 1	Trial 2	Trial 3
Number of drops of NaOH solution used			
Number of liters of NaOH solution used			
Number of moles of NaOH in sample			
Number of drops of $HC_2H_3O_2$ solution used			
Number of liters of $HC_2H_3O_2$ solution used			
Calculated molarity, M, of the $HC_2H_3O_2$ solution			
Average M of the vinegar ($HC_2H_3O_2$) solution			

Based on the average molarity, M, of the vinegar solution, if acetic acid, $HC_2H_3O_2$, has a molar mass of is 60 g/mole, how many grams of acetic acid are present in a liter of vinegar?

$$\underline{}\ \frac{\text{moles } HC_2H_3O_2}{\text{liter vinegar}} \times \frac{60 \text{ grams } HC_2H_3O_2}{\text{mole } HC_2H_3O_2} = \underline{}\ \frac{\text{g } HC_2H_3O_2}{\text{liter vinegar}}$$

5. Weight-volume percent is found by dividing the mass of solute by its volume, expressed in milliliters, times 100. Using the following formula, calculate the $\%_{(wt/vol)}$ of acetic acid in vinegar.

$$\frac{\text{g } HC_2H_3O_2}{\text{liter vinegar}} \times 100 = \text{_____ } \%_{(wt/vol)}$$

Extra Credit

If the titration of a 10.0 mL vinegar sample required 29.07 mL of standard 0.250 M NaOH, what is the molarity and weight/volume percent of acetic acid in the sample. Show your work.

Can you build your own voltmeter and use it to determine a mystery voltage?

Build Your Own Voltmeter

■ Background Concepts

The goal of this experiment is to design and build a voltmeter that can be used to measure a mystery voltage. A voltmeter is little more than a galvanometer that has the appropriate resistance in its circuit so that the potential difference between two points in a circuit produces nearly a full-scale deflection of the needle.

When a voltage is applied across a resistor, an electric current flows. Like water valves, different resistors will let different amounts of current pass. We define the resistance, R, of the resistor by how much current "I" it allows to flow. This is called Ohm's law, $I = V/R$, where V is the amount of voltage (measured in volts) applied across the resistor, I is the amount of current (measured in amperes) it allows to flow, and R is the amount of resistance (in ohms) the resistor offers.

For example, suppose a 10 Ω resistor is used to connect the two ends of a AA battery. The AA battery supplies 1.5 volts across the resistor. Ohm's law shows that the resistor will let a current of only 0.15 amps (1.5 V/10 Ω = 0.15 amps) flow through it. This means that the number of electric charges flowing through the resistor in one second would be 0.15 coulomb (1 amp = 1 coulomb/second). The 1.5 volts means that the battery gives each coulomb of charges 1.5 joules of energy to get through the circuit. The electric charges, however, must be losing this energy as they pass through the resistor, just like bullets lose kinetic energy when they pass through a solid. We describe this energy loss as the voltage "drop" across the resistor, because the electric charges coming out of the resistor have 1.5 joules per coulomb less energy than they did upon entering. We therefore say that the resistor "draws" a current of 0.15 A or 150 mA from the battery. For an electrical current to flow through a circuit there must be a continuous path, forming a closed loop. Otherwise, the electrical charges would pile up like people entering a subway that has no outlet.

There are two ways to place resisters in a circuit – in series or in parallel. When they are placed in series it means they are connected end-to-end. Let's say we connect a 10 Ω resistor to a 5 Ω resistor, and then connect the two remaining wires to the two ends of a battery. The electric charges flowing into the first (10 Ω) resistor must pass through it before they can pass through the second (5 Ω) resistor. When resistors are connected in this way, they must have the same amount of current flowing through each one, and $I_{10\Omega} = I_{5\Omega}$. In this arrangement, the electric charges must still lose energy as they flow through the two resistors – only now they will lose only part of it as they pass through the 10 Ω resistor and the remainder of it as they pass through the 5 Ω resistor. Together, they will still have a total voltage drop of 1.5 volts:

$$V_{\text{drop }10\Omega} + V_{\text{drop }5\Omega} = 1.5 \text{ volts, or}$$

$$I_{10\Omega} = I_{5\Omega}$$

Investigation 1

Objective

To measure the drop in voltage across each resistor in a series circuit and relate it to the ratio of their resistances.

Procedure

1. Your instructor will provide you with 2 resistors. Set your digital multimeter to the kilohm (K) setting (around 40 K), to measure resistance. For each resistor, touch your probes to the ends of the resistor to measure its value.

   (Convert: 1 K = 1,000 ohms)

$$R_1 = \text{_____} \text{ ohms}$$

$$R_2 = \text{_____} \text{ ohms}$$

   (Adjust your answers so that R_1 is the resistor with the smaller resistance.)

2. Connect your resistors in series: end-to-end, so that each has one free end. (This can be done by using a wire with alligator clips at both ends and clipping it between the two resistors; although this procedure may vary depending on your equipment.)

3. By touching your multimeter probes to the two free ends, measure the combined resistance of the two resistors in series:

$$R_{\text{combined}} = \text{_____} \text{ ohms}$$

4. Now set your multimeter to the DC Volts setting (around 4 V), so that you can measure voltage.

5. Your instructor will provide you with a battery. Touch your probes to both ends to determine the voltage the battery provides.

$$V_{\text{battery}} = \text{_____} \text{ volts}$$

6. Using alligator clip wires, connect the battery to the series combination of resistors to form the circuit as shown in Figure 17-1.

7. Figure 17-2 is a voltage diagram showing how the voltage drops throughout this simple circuit. Let's say that the negative end of the battery (A) is where we will consider the voltage to be zero. Connect the negative probe (the black one) to this point.

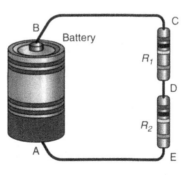

Figure 17-1

8. Now by using the red probe, you can measure the voltage at each point in the circuit, relative to the negative end of the battery. On the voltage diagram, there are 5 boxes labeled A, B, C, D, and E. Measure the value of the voltage at each of these points, and place their values in the appropriate spaces on Figure 17-2.

9. To prolong the useful life of the battery, disconnect it from the circuit.

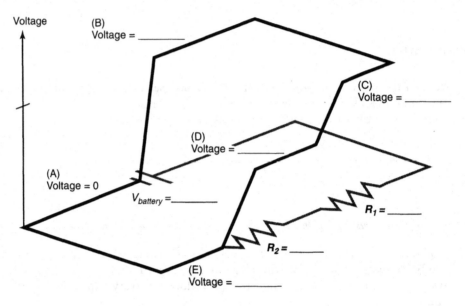

Figure 17-2

Interpretation

1. How is $R_{combined}$ related to the values of R_1 and R_2?

2. How much higher in voltage is a charge at point B than a charge at point A?

3. Does the voltage change as charges go from B to C? Explain.

4. Using Ohm's law, ($V = I R$), predict how much current (I) flows through the first resistor, R_1:

$$R_1 = \text{_____ amperes}$$

5. How much does the voltage drop as the charges go through R_1?

6. How much of a voltage drop is left for the charges to experience when they go through R_2?

7. What was the total drop in voltage from point C to point E?

8. What percentage of the total voltage drop across C–E does your answer to question 5 represent?

9. What percentage of the total combined resistance does R_1 represent?

10. Using Ohm's law, calculate the current flowing through the whole circuit:

$$I = \frac{\Delta V_{\text{total (C–E)}}}{R_{\text{combined}}} = \underline{\hspace{1.5cm}} \text{ amperes}$$

11. Using Ohm's law, calculate the current flowing through R_2:

$$I_2 = \frac{\Delta V_{2\ (\text{D–E})}}{R_2} = \underline{\hspace{1.5cm}} \text{ amperes}$$

12. Using Ohm's law, calculate the current flowing through R_1:

$$I_1 = \frac{\Delta V_{1\ (\text{C–D})}}{R_1} = \underline{\hspace{1.5cm}} \text{ amperes}$$

13. Are all three of these I's the same? Why should they be?

14. How does the result from question 12 compare with the prediction you made in question 4? If it is different, explain why your prediction was wrong.

15. Point D is roughly 1/3 of the way through the total combined resistance of the circuit. Is the voltage at that point roughly 1/3 less than at point C?

Investigation 2

Objective

To determine the characteristics of a galvanometer.

Procedure

1. Your instructor will provide you with a galvanometer. The galvanometer is very sensitive and can read both positive and negative flows of electricity. If the current is too high, it can cause the needle to break off.

2. Set your digital multimeter to the Ohms Ω setting, at its smallest scale (probably around 400 Ω). By touching its probes to the terminals of the galvanometer, determine the internal resistance of the galvanometer.

$r_1 = $ _____ ohms

3. To determine how much voltage it takes to produce a full-needle deflection of your galvanometer, your instructor will provide you with a potentiometer. A potentiometer is actually a tuning knob with three prongs, (a), (b), (c). Turn the potentiometer knob to about 1/3 of its full range. Using your digital multimeter in the kilohm (K) setting, measure the resistance between:

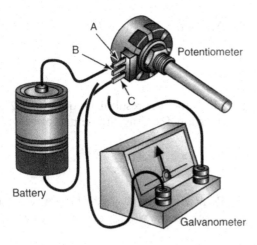

Figure 17-3

(a) and (b): _____ kilohms

(b) and (c): _____ kilohms

(a) and (c): _____ kilohms

4. The potentiometer is made of a large resistor between (a) and (c). When you turn the knob, you move the middle contact (b) along the resistor. If you originally turned the knob 1/3 of the way, you should see that the resistance between (a) and (b), is about 1/3 of the total resistance between (a) and (c). Verify that this is so.

5. Connect the (a) and (c) prongs of the potentiometer to the battery with two wires as shown in Figure 17-4.

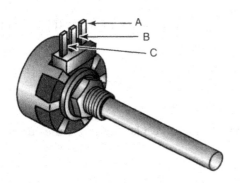

Figure 17-4

6. Connect the two wires from the galvanometer terminals to the (b) and (c) potentiometer prongs. Carefully adjust the knob on the potentiometer until the needle points to the scale's largest reading (5), without going past it. (If the needle deflects the wrong way, it may be necessary to reverse the two wires.)

7. Set your digital multimeters to DC volts. Touch your probes to the two terminals of the galvanometer to measure the voltage it takes to deflect the needle:

$$V_{max} = \underline{\hspace{2cm}} \text{ volts}$$

(*NOTE: The voltage must be measured across the galvanometer <u>while</u> the needle is being deflected.*)

8. To prolong the useful life of the battery, disconnect it from the circuit.

Interpretation

1. Let's consider the circuit in step 5 of the Procedure, before you connected the galvanometer. This circuit is similar to the one you considered in the previous investigation. To make the comparison, we would say that:

   — $R_{(a)-(b)}$ is like R_1

   — $R_{(b)-(c)}$ is like $\underline{\hspace{2cm}}$

   — $R\underline{\hspace{1.5cm}}$ is like $R_{combined}$

   — and the (b) prong is like point D.

2. As you adjust the knob on the potentiometer, you change how much of the total resistance is between (a) and (b), and how much of it is between (b) and (c). What effect does this have on the (b) prong reading?

3. Using Ohm's law, calculate how much current it takes to deflect the galvanometer needle all the way to (5). ($V_{max} = r_1 I_{max}$)

$$I_{max} = \underline{\hspace{2cm}} \text{ amperes}$$

■ Investigation 3

Objective

To determine the multiplier resistor to use in constructing your voltmeter and use it to measure a mystery voltage.

Procedure

1. We want to measure a mystery voltage with our constructed voltmeter. We know that the mystery voltage might be as high as 5 volts. However, the largest voltage drop our voltmeter can measure is:

$$V_{max} = \underline{\hspace{2cm}}$$

(from Investigation 2)

2. We will fix this by adding a resistor, R_m, to our galvanometer circuit. The question is what resistance does R_m need to be? We want it so that when a voltage of 5 volts is applied to both the resistor and galvanometer, the actual current that flows is:

$$I_{max} = \underline{\hspace{2cm}}$$

(from Investigation 2)

3. In order for that much current to flow when 5 volts is applied, what does the total combined resistance need to be?

$$V = I\,R$$

$$R_{combined} = \frac{V}{I} = \frac{5 \text{ volts}}{I_{max}} = \underline{\hspace{2cm}} \ \Omega$$

4. Calculate what R_m needs to be, in order to have that much total combined resistance:

$$R_{combined} = r_I + R_m$$

$$R_m = \underline{\hspace{2cm}} \ \Omega$$

5. Set your digital multimeter to the kilohms (K) setting (around 50 K). Disconnect the potentiometer from the circuit.

6. Connect the multimeter probes to prongs (a) and (b). Carefully adjust the potentiometer so that the resistance between these two prongs is exactly what you calculated for R_m.

7. Connect the potentiometer to the galvanometer in series, as shown in Figure 17-5. Together these form a voltmeter, since the scale measures voltage. This way, the potentiometer acts like R_2, dropping a large portion of the applied voltage. That will leave the galvanometer with a much smaller voltage drop.

8. Using the combination of potentiometer/galvanometer, measure the mystery voltage source provided by the instructor.

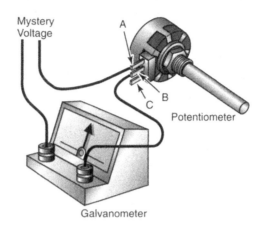

Figure 17-5

Interpretation

1. What is the value of the "mystery voltage"?

_____ volts

Report Sheet

Name: _____

Experiment 17

Date: _____ Section: _____

---

1. On Figure 17-6 (voltage diagram), fill in the values you measured in Investigation 1.

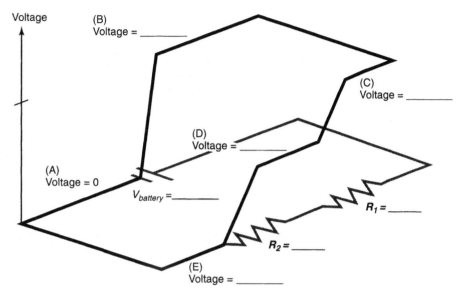

Figure 17-6

2. What percentage of the total resistance is R_1?

_____ %

3. What percentage of the total voltage drop occurs across R_1?

_____ %

4. Galvanometer characteristics:

   a. Internal resistance: _____ ohms

   b. Full-scale voltage: _____ volts

   c. Full-scale current: _____ amperes

5. Mystery voltage

_____ volts

6. This was measured using an additional resistance with a value of:

R_m = _____

7. Why must two resistors, connected in series, necessarily have the same current flowing through them?

8. In Investigation 1, what would have been the voltage at (D) if the larger of the two resistors had been used as R_1, and the smaller as R_2.

9. In Investigation 2, suppose you connect the (a) and (c) prongs of the potentiometer to the battery, and you tune the knob to the middle of the range. What would be the voltage between prongs (a) and (b)? What would be the voltage between prongs (b) and (c)?

10. Why do you suppose the potentiometer in Investigation 2 is sometimes called a "voltage divider"? Explain.

11. In Investigation 3, you built a voltmeter that could read up to 5 volts. Suppose you needed it to read as high as 50 volts, using the same galvanometer. What value of R_m would you use?

Extra Credit

1. Often, instead of using the exact R_m to make a voltmeter, we round it up. So, if R_m needed to be just less than 10 K, we would use a 10 K resistor anyway. If you did that with your galvanometer and measured what was really a 3 V mystery voltage, what would the galvanometer scale actually read?

Can you build your own ammeter and use it to determine a mystery current?

Build Your Own Ammeter

Background Concepts

An ammeter is an instrument used to measure, in amperes, the magnitude of an electric current. In this experiment, you will be given a circuit with a mystery current flowing through it. Your task is to design and build an ammeter capable of measuring it, and to determine the value of the mystery current.

When a voltage is applied across a resistor, an electric current flows. Like water valves, different resistors will let different amounts of current pass. We define the resistance, R, of the resistor by how much current, I, it allows to flow. This is called Ohm's law, $I = V/R$, where V is the amount of voltage (measured in volts) applied across the resistor, I is the amount of current (measured in amperes) it allows to flow, and R is the amount of resistance (in ohms) the resistor offers.

For example, suppose a 10 Ω resistor is used to connect the two ends of a AA battery. The AA battery supplies 1.5 volts across the resistor. Ohm's law shows that the resistor will let a current of only 0.15 amps (1.5 V/10 Ω = 0.15 amps) flow through it. This means that the number of electric charges flowing through the resistor in one second would be 0.15 coulomb (1 amp = 1 coulomb/second). The 1.5 volts mean that the battery gives each coulomb of charges 1.5 joules of energy to get through the circuit. The electric charges, however, must be losing this energy as they pass through the resistor, just like bullets lose kinetic energy when they pass through a solid. We describe this energy loss as the voltage "drop" across the resistor, because the electric charges coming out of the resistor have 1.5 joules per coulomb less energy than they did upon entering. We therefore say that the resistor "draws" a current of 0.15 A or 150 mA from the battery.

For an electrical current to flow through a circuit there must be a continuous path, forming a closed loop. Otherwise, the electrical charges would pile up like people entering a subway that has no outlet. Connecting two resistors "in parallel" means connecting their two front ends together, and connecting their two back ends together, in a circuit. When the charges reach their front ends, some of the charges can go through one resistor, and the rest of them can go through the other.

Let's say we connect a 10 Ω and a 5 Ω resistor in parallel. The resistors now share common front and back ends, which are connected to the terminals of a 1.5 volt battery. The electric charges entering must now choose which resistor to go through. Whether they go through the 10 Ω resistor or the 5 Ω resistor, a closed loop is formed with the battery.

The 10 Ω resistor will allow a current of (1.5 V/10 Ω =) 0.15 amps to pass through it, while the 5 Ω resistor will allow a current of (1.5 V/5 Ω =) 0.3 amps to flow through it. Together, the two resistors "draw" a combined current of (0.15 amps + 0.3 amps =) 0.45 amps from the battery.

The electric charges must still lose their energy as they flow through one of the two resistors – some will lose it passing through the 10 Ω resistor and the others will lose it passing through the 5 Ω resistor. Both sides will experience the same voltage drop of 1.5 volts, even though more current is flowing through one resistor than the other:

$$V_{drop\ 10\Omega} = 1.5 \text{ volts}$$

$$V_{drop\ 5\Omega} = 1.5 \text{ volts}$$

$$V_{drop\ 10\Omega} = V_{drop\ 5\Omega},$$

$$I_{10\Omega} + I_{5\Omega} = I_{total}.$$

Investigation 1

Objective

To measure the voltage drop across each resistor in a parallel circuit and relate it to the ratio of their resistances.

Procedure

1. Your instructor will provide you with 2 resistors. Set your digital multimeter to the "kilohm" (K) setting (around 40 K), to measure resistance. For each resistor, touch your probes to both ends of the resistor to measure its value.

   (Convert: 1 K = 1,000 ohms)

   $$R_1 = \underline{\hspace{2cm}} \text{ ohms}$$

   $$R_2 = \underline{\hspace{2cm}} \text{ ohms}$$

   (Adjust your answers so that R_1 is the resistor with the smaller resistance.)

2. Connect your resistors in parallel, by connecting one end of one resistor to one end of the other resistor and connect their other two ends together. (Depending on your equipment, this can be done by using a wire with alligator clips at both ends, and clipping it between one resistor and the other). Now connect a wire to each of the two ends (B and E in Figure 18-1) where the resistors come together.

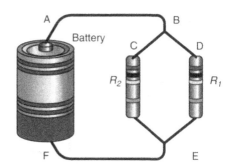

Figure 18-1

3. By touching your probes to the two free ends, measure the combined resistance of the two resistors in parallel.

   $$R_{\text{combined}} = \underline{\hspace{2cm}} \text{ ohms}$$

4. Now set your multimeter to the "DC Volts" setting (around 4 V), so that you can measure voltage.

5. Your instructor will provide you with a common battery. Touch your probes to both ends (A and F in Figure 18-1) to determine the voltage the battery provides.

   $$V_{\text{battery}} = \underline{\hspace{2cm}} \text{ volts}$$

6. Using alligator clip wires, connect the battery to the series combination of the two parallel resistors to form the circuit shown in Figure 18-1.

7. Figure 18-2 is a voltage diagram showing how the voltage drops throughout this simple circuit. Fill in the line for how high the battery raises the voltage.

8. Set your multimeter to the "DC – A" setting, to measure current. Choose a scale of about 400 mA. For many multimeters, using this setting will require you to unplug the red probe, and connect it to a different jack on the multimeter, which

may be labeled "A" or "400 mA". Verify this with your instructor if the labeling is not clear.

9. Using your multimeter, measure the current flowing into the resistors. To do this, disconnect the wire between point A and point B (this will break the circuit and stop the flow). Touch your negative (black) probe lead to point A and your positive (red) probe lead to point B (this will resume the flow by allowing it to flow through the multimeter). Measure the current passing through the multimeter, and record it on the diagram as I_{in}. If necessary, adjust the scale on the multimeter to a more appropriate scale.

10. Remove the multimeter from the circuit, and reconnect the circuit the way it was before step 9. Repeat step 9 to find the current going through R_1 by "inserting" the multimeter between point B and point D. Record this as I_1 on Figure 18-2. Repeat the procedure to find I_2 (inserting the multimeter between B and C) and I_{out} (between E and F).

11. To prolong the useful life of the battery, disconnect it from the circuit.

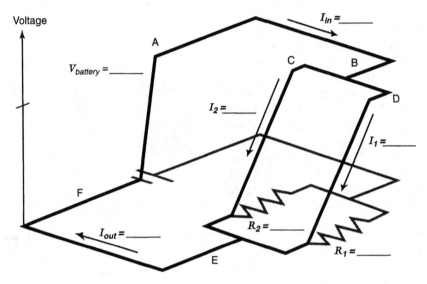

Figure 18-2

Interpretation

1. Compute the following values:

$$\frac{1}{R_1} + \frac{1}{R_2} = \underline{\hspace{2cm}} \ \Omega^{-1}$$

$$\frac{1}{R_{combined}} = \underline{\hspace{2cm}} \ \Omega^{-1}$$

Are they nearly equal?

2. Is I_{in} equal to I_{out}? Why should it be?

3. Is I_{in} equal to $I_1 + I_2$? Why should it be?

4. Using Ohm's law, what should be the voltage drop across the first resistor?

$$V_{drop} = I_1 \times R_1 = \underline{\hspace{1.5cm}} \text{ volts}$$

5. Predict what the voltage drop should be across the second resistor, R_2.

6. Using Ohm's law, calculate the voltage drop across the second resistor.

$$V_{drop} = I_2 \times R_2 = \underline{\hspace{1.5cm}} \text{ volts}$$

7. Are these two voltage drops equal? Why should they be?

8. By what factor is R_2 larger than R_1?

$$R_2 \text{ is } \underline{\hspace{1.5cm}} \text{ times as large as } R_1$$

9. By what factor is I_1 larger than I_2?

$$I_1 \text{ is } \underline{\hspace{1.5cm}} \text{ times as large as } I_2$$

10. Why is it in this case that adding two resistors together actually gives you a smaller combined resistance?

11. How much power does each resistor dissipate?

$$Power_1 = V_{drop} I_1 = \underline{\hspace{1.5cm}} \text{ watts}$$

$$Power_2 = V_{drop} I_2 = \underline{\hspace{1.5cm}} \text{ watts}$$

12. If the smaller resistor is easier to flow through, why does it dissipate more power as heat?

Physical Science: What the Technology Professional Needs to Know　　　　　　　**239**

Investigation 2

Objective

To determine the characteristics of a galvanometer.

Procedure

1. Your instructor will provide you a galvanometer. The galvanometer is very sensitive and can read both positive and negative flows of electricity. If the current is too large, it can cause the needle to break off.

2. Set your digital multimeter to the "Ohms Ω" setting, at the smallest scale (probably around 400 Ω). By touching your probes to the terminals of the galvanometer, determine the internal resistance of the galvanometer.

$$r_{\mathrm{I}} = \underline{\hspace{2cm}} \text{ ohms}$$

3. To determine how much voltage it takes to produce a full-needle deflection of your galvanometer, your instructor will provide you a potentiometer. A potentiometer is actually a tuning knob with three prongs, (a), (b), (c) as shown in Figure 18-3. Set the knob roughly 1/3 of it its full range. Using your digital multimeter in the "kilohm (K)" setting, measure the resistance between:

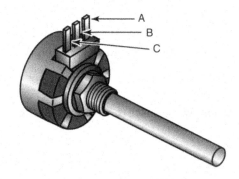

(a) and (b): _____ kilohms

(b) and (c): _____ kilohms

Figure 18-3

(a) and (c): _____ kilohms

4. The potentiometer is made of a large resistor between (a) and (c). When you turn the knob, you move the middle contact (b) along the resistor. If you are turned the knob 1/3 of the way, you should see that the resistance between (a) and (b), is 1/3 of the total resistance between (a) and (c). Verify that this is so.

We will use only the (b) and (c) prongs as an adjustable resistor. Connect the potentiometer to the battery and galvanometer as shown in Figure 18-4.

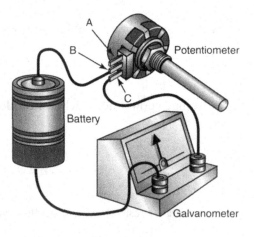

Figure 18-4

5. Carefully adjust the knob on the potentiometer until the galvanometer's needle points to the scale's largest reading (5), without going past it. (If the needle deflects the wrong way, it may be necessary to reverse the two wires.)

6. Set your digital multimeters to "DC volts" and a scale of a few hundred mV. (You may need to reconnect the red probe lead to the appropriate jack of the multimeter). Touch your probes to the two terminals of the galvanometer to measure the voltage it takes to deflect the needle:

$$V_{max} = \underline{\hspace{2cm}} \text{ volts}$$

(*NOTE: The voltage must be measured across the galvanometer <u>while</u> the needle is being deflected.*)

7. To prolong the useful life of the battery, disconnect it from the circuit.

Interpretation

1. Is the "variable resistor" connected to the galvanometer in parallel, or in series?

2. As you adjust the knob on the potentiometer, you change the amount of resistance between (b) and (c). How does this change the total combined resistance?

3. How does it affect the current flowing through the galvanometer?

4. Using Ohm's law, how much current does it take to deflect the galvanometer's needle all the way to (5)? ($V_{max} = r_I\, I_{max}$)

$$I_{max} = \underline{\hspace{2cm}} \text{ amperes}$$

Investigation 3

Objective

To determine what shunt resistor to use in constructing your ammeter and then use it to measure a mystery current.

Procedure

1. We want to measure a mystery current with our constructed ammeter. We know that the mystery current might be as high as 50 mA (0.050 amperes). However, the largest current our ammeter can register is:

   $$I_{max} = \underline{\hspace{2cm}}$$
   (from Investigation 2)

2. We will fix this by adding a "shunt" resistor "R_s" in parallel to our galvanometer, to deviate most of the current. That way, only a fraction of the current we're trying to measure will flow through the galvanometer. This is similar to Investigation 1, where only a small fraction of the whole current (I_{in}) flowed through R_2.

   In terms of Investigation 1, we want our galvanometer to be like R_2, and our shunt resistor to be like

   _____

3. The question is: how big does R_s need to be? We want it so that when a current of 50 mA is coming in as our I_{in}, the actual current that flows through the galvanometer is I_{max}.

   $$I_{max} = \underline{\hspace{2cm}}$$
   (from Investigation 2)

4. In order for that much current to flow through the galvanometer when our I_{in} is 0.050 amperes, how much current is going to be left that needs to flow through R_s?

   $$I_s = \underline{\hspace{2cm}} \text{ amperes}$$

5. How many times bigger is I_s than the galvanometer's I_{max}?

   $$I_s \text{ is } \underline{\hspace{1.5cm}} \text{ times as large}$$

6. Based on the Interpretation of Investigation 1, how much smaller should R_s be, compared to the galvanometer's internal resistance, r_I?

   $$R_s \text{ needs to be } \underline{\hspace{2cm}} \text{ times as small as } r_I$$

   $$\text{Therefore, } R_s = \underline{\hspace{2cm}} \text{ ohms}$$

7. Let's test it and see.

   a. What would be the voltage drop across the galvanometer?

   $$V_{drop} = r_I \times I_{max} = \underline{\hspace{2cm}} \text{ volts}$$

   b. What would be the voltage drop across the shunt resistor, R_s?

   $$V_{drop} = R_s \times I_s = \underline{\hspace{2cm}} \text{ volts}$$

   c. Do both of these work out to the same voltage drop?

   _____

8. You will be provided with a 5 ohm potentiometer. Set your digital multimeter to the lowest "ohms (Ω)" setting.

9. Connect your two probes to the prongs (b) and (c). Carefully adjust the potentiometer so that the resistance between these two prongs is exactly what you calculated for R_s.

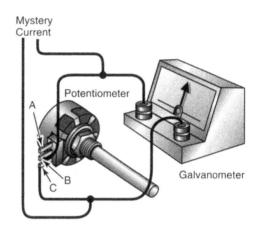

Mystery Current

Potentiometer

A

B

C

Galvanometer

Figure 18-5

10. Connect the potentiometer to the galvanometer in parallel, as shown in Figure 18-5. This way, the potentiometer acts as R_S, deviating a large portion of the total current. That will leave a much smaller current to flow through the galvanometer. Together, these form a simple ammeter, with which you can measure current.

_____ mA

11. Using the combination of potentiometer/galvanometer, measure the "mystery current source" provided by the instructor. To do this, you will have to "break" the mystery circuit (i.e. disconnect one of its wires), and insert your ammeter as part of the circuit as shown. That's the only way to get the current in the circuit to flow through your ammeter. (Note: there may be a switch you have to activate to turn the mystery circuit on.)

Interpretation

1. What is the value of the "mystery current"?

_____ amperes

Report Sheet

Name: _____

Experiment 18

Date: _____ **Section:** _____

---

1. On Figure 18-6 (voltage diagram), fill in the values you measured from Investigation 1.

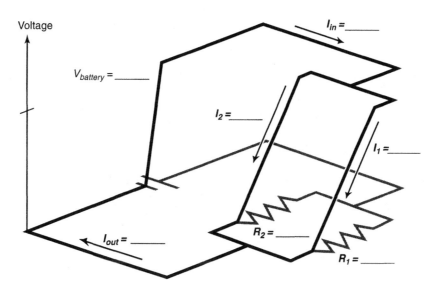

Figure 18-6

2. What percentage of the total current flows through R_1?

_____ %

3. What percentage of the total current flows through R_2?

_____ %

4. Combined total resistance of R_1 and R_2 in parallel:

_____ ohms

5. Galvanometer characteristics:

   a. Internal resistance:

   _____ ohms

   b. Full-scale voltage:

   _____ volts

   c. Full-scale current:

   _____ amperes

6. Mystery current:

   _____ amperes

7. This was measured using a shunt resistance with a value of:

   R_s = _____

8. In this lab you built an ammeter that could read up to 50 mA. Suppose instead you needed it to read as high as 0.5 amperes, using the same galvanometer. What value of R_s would you use?

9. What is the combined resistance of your R_s and the galvanometer's r_I? This is the "internal" resistance of your ammeter, $r_{ammeter}$.

10. When two resistors are connected in parallel, which one dissipates the most power?

Extra Credit

1. Suppose that your "mystery current source" was produced by a 9 volt battery connected through some resistor, R_{source}.

   a. How big would R_{source} have to be to produce the current you measured?

   b. How much would you be changing the current by adding or removing your ammeter, (therefore adding or removing an extra resistance $r_{ammeter}$ from the 9 volt battery?

What is the index of refraction of acrylic?

Refraction

Background Concepts

When a ray of light leaves one medium and enters another, its direction is nearly always changed. This bending of the light ray is called refraction and is the result of the light having a slightly different velocity in one medium than another. It is described by Snell's law, $n_1 \sin \theta_1 = n_2 \sin \theta_2$, where θ_1 is the angle of the ray in the first medium and θ_2 is the angle of the ray in the second medium. The angles are measured from the normal or "head-on" direction. A beam that's at zero degrees, $0°$, is entering the medium perpendicular to its surface and a beam that's at $90°$ would be skimming the surface of the medium. The "n" in the equation is called the "index of refraction," which characterizes how light gets bent upon entering that medium.

By definition, the index of refraction for empty space (a vacuum) is 1.00. It's just a bit more for air: $n_{air} = 1.0003$. For water it's even more: $n_{water} = 1.33$. Therefore, we say that water is more *optically dense* because it has a higher index of refraction than air. Actually, it turns out that a material's index of refraction is related to how fast light can travel in that material:

$$n_{material} = \frac{\text{speed of light in vacuum}}{\text{speed of light in the material}}$$

Because light travels more slowly (smaller number) in water than it does in a vacuum (larger number), the value of n for water is more than 1.

If a beam of light comes out of water and enters the air, then rearrangement of Snell's law produces:

$$\sin \theta_{air} = \frac{n_{water}}{n_{air}} \sin \theta_{water}$$

From this equation, it can be seen that if n_{water} divided by n_{air} is greater than 1 and the value of $\sin \theta_{water}$ increased, a point will be reached where their product is greater than 1. A sine function, such as $\sin \theta_{air}$, can never be greater than 1, however, so at this angle there must be no refraction at all. In other words, there is no possible mathematical solution for θ_{air}, and we find that the beam does not enter the air at all, but rather all of it is reflected back into the water. Note that reflection always occurs to some extent, even at angles where there is a mathematical solution for θ_{air}. The difference here is that when there is no mathematical solution for θ_{air} the entire beam is reflected, since none of it can be refracted.

The largest angle that a light beam in water can have and still refract into the air can be calculated by knowing the respective indices of refraction. If the angle exceeds the critical angle, $\theta_{critical}$, then the beam will only be reflected back into the water. The equation for calculating the critical angle is:

$$\theta_{critical} = \sin^{-1}\left(\frac{n_{air}}{n_{water}}\right)$$

In this experiment you will use a laser to investigate refraction and then use the results to calculate the index of refraction of the two materials: water and acrylic.

Investigation 1

Objectives

To calculate index of refraction of water based on its angle of refraction and critical angle.

Procedure

1. You will be provided with a laser and a semi-circular tank.

> **Note:** When handling a laser, be careful to NOT shine it into anyone's eyes.

   Remove the protractor sheet from the end of this experiment. Fill the tank with DI water. Center the tank on the protractor sheet. Orient the tank on the sheet as shown. The 90° line should be aligned with the inside edge of the tank.

2. As shown in Figure 19-1, align the laser so the beam enters the round side of the tank and passes through the center. This should correspond to the center of curvature of the tank, and the center of the protractor sheet. Lab jacks may be used to align the laser with the tank.

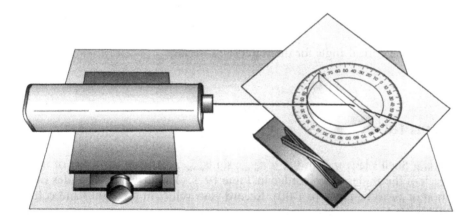

Figure 19-1

3. The laser beam should exit from the center of the flat side of the tank. Note the angle at which the laser beam exits the flat side of the tank. This is the angle θ_{air}. Also locate the angle from which the laser entered the water on the round side. This is the angle θ_{water}, and is the angle at which the laser beam in the water strikes the flat side of the tank at the center of the protractor sheet.

4. Now, set the laser so the beam enters at $\theta_{water} = 10°$. Locate the angle at which the beam emerges on the flat side. Record this angle as θ_{air} in Table 19-1. Repeat this procedure for the other values of θ_{water} listed in the table, and record the value of θ_{air} where the beam emerges. In each case, verify that the beam is emerging from the center point of the protractor sheet.

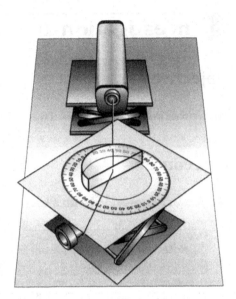

Figure 19-2

θ_{water} (degrees)	θ_{air} (degrees)	n_{water}
10		
20		
30		
40		

Table 19–1: Water/air laser beam angles.

5. Gradually increase the angle θ_{water} for the laser beam until it no longer emerges on the flat side of the tank, or if it does, only at a 90° angle. What is the largest value of θ_{water} in which the beam emerges on the flat side of the tank and at an angle of less than 90°?

This is the critical angle for the water/air interface:

$$\theta_{critical} = \underline{\hspace{2cm}} \text{ degrees}$$

Interpretation

1. Using Snell's law, $n_{air} \sin \theta_{air} = n_{water} \sin \theta_{water}$, calculate the value of n_{water} for each of the angles you recorded in Table 19-1. (You may take the index of refraction of air to be equal to 1.00). Record your values in the right hand column of Table 19-1.

2. What is the average value of n_{water}?

$$n_{water} = \underline{\hspace{2cm}}$$

3. Using the formula for the critical angle, $\sin \theta_{critical} = n_{air} / n_{water}$ and your measurement of the critical angle, calculate the index of refraction of water.

Investigation 2

Objectives

To sketch the path of an object entering an area where its speed changes.

To relate refraction to the changing speed of light as it crosses from one medium to another.

Procedure

1. Figure 19-3 depicts a figure of a car. The car has front wheel drive. Let's say that the car is driving on regular pavement as it crosses into a patch of mud. Because of the angle of the car, the right front tire will encounter the mud first. Sketch the direction that the car is likely to take as it enters the muddy patch.

Figure 19-3

2. Explain why you drew the path the way you did.

3. Now imagine a similar situation, only the car is in the mud and slowly making its way out onto the pavement as shown in Figure 19-4. Sketch the path the car is likely to follow as it comes onto the pavement.

Figure 19-4

4. Explain why you drew the path the way you did.

Interpretation

1. The speed of light in water is slower than the speed of light in air. If you picture the laser beam as a wide, straight wave front moving toward the interface, how can you explain why the wave front changes direction as it crosses the interface from water to air? Which of these two car scenarios does it most resemble?

Investigation 3

Objectives

To calculate index of refraction of acrylic plastic based on its angle of refraction and critical angle.

Procedure

1. You will be provided with a semicircular acrylic block. Place it on the protractor sheet as you did the water tank in Investigation 1. Take care to align the outside edge of the flat side of the block with the 90° line on the sheet. The center of the protractor sheet should coincide with the center of curvature of the block.

2. Repeat the measurements you made in Investigation 1. Point the laser at an angle $\theta_{acrylic}$ into the round side, so that it crosses the center of the protractor sheet at this angle and is refracted out the flat side, at an angle θ_{air}. Locate the angle at which the laser is refracted out of the block. Record your measurements for each of the angles in the Table 19-2.

$\theta_{acrylic}$ (degrees)	θ_{air} (degrees)	$n_{acrylic}$
10		
20		
30		
40		

Table 19–2: Acrylic/air laser beam angles.

3. Gradually increase the angle $\theta_{acrylic}$ at which the laser beam is directed into the block until the beam no longer emerges on the flat side. What is the largest value of $\theta_{acrylic}$ for which the beam emerges on the flat side at an angle of less than 90°?

   This is the critical angle for acrylic/air:

   $$\theta_{critical} = \underline{\hspace{1cm}} \text{ degrees}$$

Interpretation

1. Using the refraction formula, $n_{air} \sin \theta_{air} = n_{acrylic} \sin \theta_{acrylic}$, calculate the value of $n_{acrylic}$ from each of the angles you measured and recorded in Table 19-2. (As in Investigation 1, you may assume the index of refraction of air is 1.00). Record your calculated values in the right-hand column of Table 19-2.

2. What is the average value of $n_{acrylic}$?

$$n_{acrylic} = \underline{\hspace{2cm}}$$

3. From the formula for the critical angle, $\sin \theta_{critical} = n_{air}/n_{acrylic}$ and your measurement of the critical angle, calculate the index of refraction for acrylic plastic.

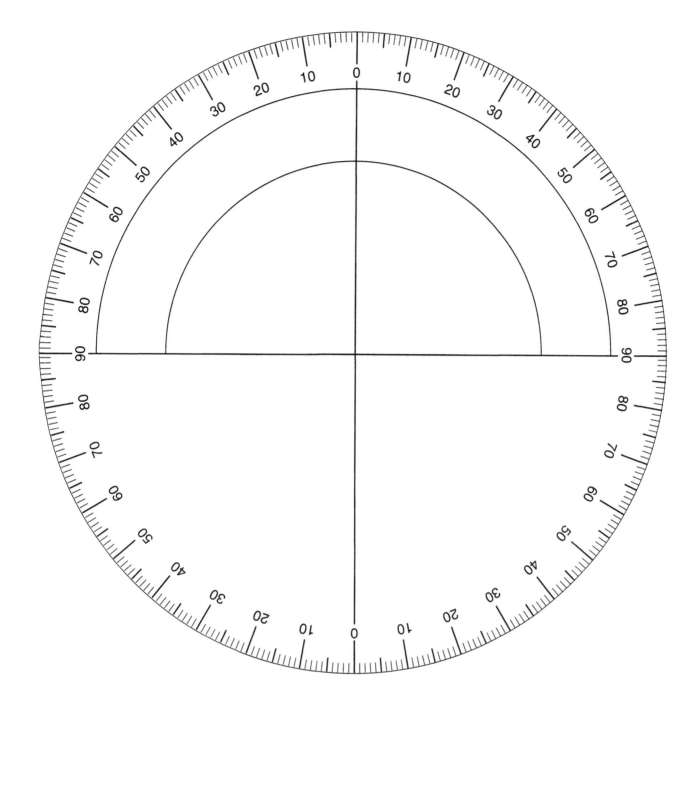

Report Sheet

Experiment 19

Name: _____

Date: _____ Section: _____

The index of refraction of distilled/DI water is 1.33.

1. What is the refractive index of water:

   a. as determined by your data in Investigation 1?

n_{water} = _____

   b. as determined by the critical angle from Investigation 1?

n_{water} = _____

2. Which method gave the more accurate answer?

3. Calculate the percent deviation for this method:

_____ %

4. What is the refractive index of acrylic:

   a. as determined by your data in Investigation 3?

$n_{acrylic}$ = _____

   b. as determined by the critical angle from Investigation 3?

$n_{acrylic}$ = _____

   c. Which of these results do you trust more?

   d. Why did you choose this method?

5. Different kinds of glass have different indices of refraction. The glass used in some eyeglasses has an index of refraction of 1.66 (these lenses are called "high-index" lenses). What difference would it make if the lenses were made of acrylic instead of glass? What if they were made of diamond ($n = 2.5$)?

6. For what angle of incidence does a beam of light not get deviated?

7. Why is there no refraction at the point where the laser enters the round side of the tank or the block?

8. What would happen if the acrylic block had rough edges instead of smooth edges?

9. In fiber optics, a laser beam is used to send a signal through an optical fiber. The beam is constrained to stay inside the optical fiber because of total internal reflection. If the fiber has a refractive index of 1.5, what is the sharpest angle at which the light can bounce around inside the fiber, without any of it being lost?

10. Why is it impossible to have total reflection of a light beam traveling *in air* and striking a water surface?

Extra Credit

1. Dispersion is the phenomenon whereby the index of refraction of a material is slightly different for light rays with different wavelengths. Because of this, we see many different colors in a diamond, although it is being illuminated by white light. Explain how this happens.

How do we know the elemental composition of the Sun?

Diffraction Gratings

Background Concepts

Light is an electromagnetic wave. Just as the pieces of a rope oscillate back and forth as a wave passes by, the electric and magnetic fields of an electromagnetic wave oscillate from intense to weak. Different colors in the spectrum correspond to waves with different frequencies, f, and have different wavelengths, λ. Here is an approximate list of visible colors and their corresponding wavelengths in a vacuum.

Color	Wavelength	Color	Wavelength
Red	650 – 700 nm	Green	500 – 550 nm
Orange	600 – 650 nm	Blue	470 – 500 nm
Yellow	560 – 600 nm	Violet	400 – 470 nm

As shown in Figure 20-1, when a straight, continuous wave strikes and passes through a small opening in a barrier, the small section of the wavefront that passes through starts to spread out, causing semicircular ripples to form. That's because the piece no longer has the rest of the wavefront beside it. This spreading out of the wave is called *diffraction*. The same thing happens to light when it passes through a small opening.

Whenever two waves meet at some point, there is a question of whether they will be in phase or out of phase. If the peaks and valleys of one wave exactly match the peaks and valleys of the other, then the waves are in phase and the os-

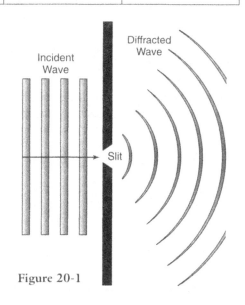

Incident Wave

Diffracted Wave

Slit

Figure 20-1

---

cillations will be stronger at that spot. This is called *constructive interference*.

On the other hand, if the waves are somewhat out of phase, then there will be some interference between them, and the spot will experience weak oscillations. If the two waves have the same exact wavelength and are completely out of phase, then the peak of one wave will be canceled by the valley of the other and there will be no oscillation at this point. This is called *destructive interference*.

A pure light wave has only one frequency, wavelength, or color. If you send a pure light wave through two or more slits, the diffracting waves that pass will each have the same wavelength, frequency, and phase. Because the light waves from each of these slits all have the same wavelength, they are able to interfere with each other, and when out of phase, can cancel each other out. When they are in phase they can reinforce each other.

A transmission diffraction grating is a slide that has a plastic film containing a series of evenly spaced slits that allow light to pass through. When you shine a pure light wave through it, many diffracting waves emerge and spread out, all having the same exact wavelength, and all starting with the same phase. At some angles, the waves interfere constructively and together they can cast a bright spot on a screen, called a *bright fringe*. The formula for finding where the bright fringe will appear is:

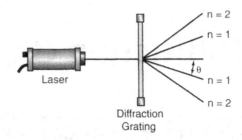

Figure 20-2

$$n\lambda = d \sin \theta, \text{ for } n = 0, 1, 2, 3 \ldots$$

Here, d is the distance between the slits, λ is the wavelength of the light, and θ is the angle where the bright fringe appears. The n stands for "1" for the first bright fringe, "2" for the second bright fringe, and so forth.

The wavefront that emerges from the slits in any particular direction is made up of rays diffracted from each slit. Each diffracted ray has to go a little bit further than the ray coming from the next slit, in order to contribute to the wavefront. This extra distance is $d \sin \theta$. If this extra distance is exactly one wavelength, 1λ, then the two rays will still be in phase where they meet and they can interfere constructively. This also works if the extra distance is exactly two wavelengths, 2λ, or exactly three wavelengths, etc. But if the extra distance isn't a whole number of wavelengths, then the rays will not be in phase as they meet, causing destructive interference. Then the light will quickly die out and no light will be seen.

So, we only see bright spots when:

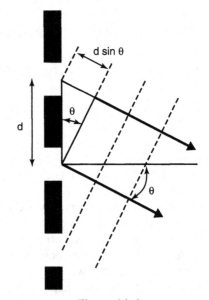

$$d \sin \theta = 1 \lambda$$
$$d \sin \theta = 2 \lambda$$
$$d \sin \theta = 3 \lambda, \text{ (etc.)}$$

Figure 20-3

This experiment demonstrates how diffraction and interference are applied to spectroscopy. The ultimate goal is to measure the bright-line emission spectrum of the element hydrogen.

Investigation 1

Objective

To determine the wavelength of a laser beam from its interference pattern.

Procedure

1. Obtain a diffraction grating from your instructor, as well as a laser.

> **NOTE:** When handling a laser, be careful to NOT shine it into anyone's eyes.

2. If you are using an optics bench, mount a screen at one end of the bench. Place the diffraction grating in a lens holder and place it on the bench about 30 cm from the screen.

3. If you are using a lab stand, secure the diffraction grating to the lab stand using a pendulum clamp as shown in Figure 20-4. Attach a sheet of paper to a bookend or tape it to the side of a box, to form a screen for the laser beam to strike.

4. Record how many lines per millimeter are engraved on the grating. (This should be written on the grating).

lines / mm = _____

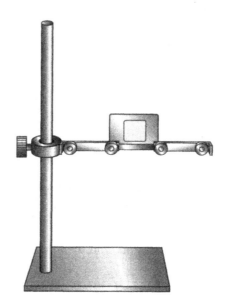

Figure 20-4: Diffraction grating.

5. Turn the laser on and point it so that the light beam passes through the diffraction grating and casts a pattern of spots on the screen. The pattern should include at least 5 spots: a bright spot in the center and 2 spots to either side of the center. It may be necessary to darken the room to see the outer most spots. Some of the side spots may also be beyond the edge of the screen.

    The first spots from the middle bright spot on either side are called the first order fringes ($n = 1$). The second spots out from the middle on either side are called the second order fringes ($n = 2$). There may be more ($n = 3, 4 \dots$) depending on your grating and the brightness of your laser beam.

6. Adjust the distance between the screen and the grating, so the middle three spots are as far apart on the screen as practical, but you still have all three on the screen.

The other two spots ($n = 2$) may disappear off the edge of the screen for now. On both sides, measure and record how far each $n = 1$ spot is from the middle spot.

$$\Delta x_1 = \text{_____ cm}$$

$$\Delta x_2 = \text{_____ cm}$$

$$\text{average: } \Delta x = \text{_____ cm}$$

7. Measure the distance between the diffraction grating and the screen:

$$L = \text{_____ cm}$$

8. Readjust the screen and diffraction grating, so that all five spots are now visible. Measure the distance between each $n = 2$ spot and the middle spot. Also measure the distance between the grating and the screen.

$$\Delta x_1 = \text{_____ cm}$$

$$\Delta x_2 = \text{_____ cm}$$

$$\text{average: } \Delta x = \text{_____ cm}$$

$$L = \text{_____ cm}$$

Interpretation

1. Based on the number of lines per millimeter for your diffraction grating, calculate the distance between lines on your grating. Express you answer in meters.

$$d = \frac{1 \text{ mm}}{\text{number of lines/mm}} = \text{_____ mm} = \text{_____ meters}$$

2. From geometry, as shown in Figure 20-5, the angle $\theta = \tan^{-1} (x / L)$.

Using this, calculate the angle for the $n = 1$ spots, and the $n = 2$ spots:

$$\theta_1 = \text{_____ degrees}$$

$$\theta_2 = \text{_____ degrees}$$

3. The condition for having a bright spot is:

$$n\lambda = d \sin \theta$$

Using your value for d, calculate what the wavelength λ is, from $n = 1$ and $\theta = \theta_1$.

Also calculate it for $n = 2$ and $\theta = \theta_2$.

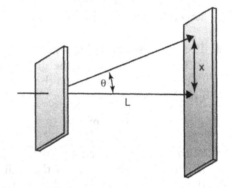

Figure 20-5: Geometry of grating.

$$\text{Using } n = 1, \lambda = \text{_____ meters}$$

$$\text{Using } n = 2, \lambda = \text{_____ meters}$$

4. According to the table in the Background Concepts section, to what color does this wavelength correspond?

■ Investigation 2

Objectives

To visualize how the interference between two sources can create bright and dark fringes.

To use the bright and dark fringes to calculate the wavelength of a directly measurable macroscopic wave.

Procedure

1. At the back of this experiment is a sheet with circular wave crests. Either your instructor will provide you with a transparent acetate photocopy of the sheet or you will need to make one.

2. On the sheet, the dark ridges represent the wave crests, while the "white" ridges represent the troughs. Using a ruler, measure and record the distance between the middle of one crest and the middle of another 10 crests away.

distance for 10 crests = _____ cm

Divide this by 10 to get the distance between successive crests. This will be your wavelength.

$\lambda_{measured}$ = _____ cm

3. As shown in Figure 20-6, place the transparent acetate over the original sheet, in such a way that the two wave sources, d, are between 1 and 5 cm apart along the wave source edge of the sheets.

4. What pattern(s) can you distinguish in the overlap of the two sets of wave crests?

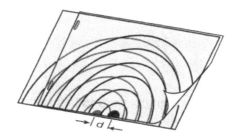

Figure 20-6: Wave overlay sheets.

5. How do these patterns change as you move the centers of the two wave sources closer together or further apart?

6. Staple the transparent acetate over the original sheet, so that the two wave sources are about 3 cm apart. Actual distance between centers:

d = _____ cm

7. Follow the center of the black and white "rays" that are formed by the two sets of crests. By peeling back the transparency, carefully mark on the bottom sheet where the middle three of these "rays" reaches the edge of the bottom sheet. Your marks should lie along the middle of the black and white rays. Also mark on the bottom sheet where the center wave source of the top transparency is located.

---

8. How far apart, along the edge of the sheets, are the marks you made for:

   a. the left ray and the middle ray?

   $$x_{1A} = \underline{\hspace{2cm}} \text{ cm}$$

   b. the right ray and the middle ray?

   $$x_{1B} = \underline{\hspace{2cm}} \text{ cm}$$

   $$\text{average: } x = \underline{\hspace{2cm}} \text{ cm}$$

9. How far away is the far edge of the paper from either of the wave source centers ?

   $$L = \underline{\hspace{2cm}} \text{ cm}$$

Interpretation

1. At what angle do the two side "rays" appear to spread out?

   $$\theta = \tan^{-1}(x / L) = \underline{\hspace{2cm}} \text{ degrees}$$

2. Since these rays correspond to $n = 1$, the wavelength should be $(1)\lambda = d \sin \theta$. Calculate the wavelength:

   $$\lambda_{calculated} = d \sin \theta = \underline{\hspace{2cm}} \text{ cm}$$

3. How well does this value match the wavelength you measured in step 2 of the procedure? Express the difference as a percent deviation:

   $$\% \text{ deviation} = \frac{\left|\lambda_{calculated} - \lambda_{measured}\right|}{\lambda_{measured}} \times 100\% = \underline{\hspace{2cm}} \%$$

4. Diffraction patterns can be hard to see when the bright spots are too close. Based on what you saw when you varied the distance between the two sources, we can make the rays spread out more by changing d. To make them spread out, do we need a larger d or a smaller d between the wave sources?

Investigation 3

Objective

To measure the wavelengths of three spectral lines of a hydrogen gas lamp.
To observe the emission spectrum of a mercury (or other) gas lamp.

Procedure

1. If possible, dim the lights in the laboratory or classroom for this activity.

2. You will be provided with a hydrogen discharge lamp. Set it at the end of your table, as shown in Figure 20-7. Place a diffraction grating in a support stand as shown, at the opposite end of the table. The grating should be oriented so that its slits are vertical. This way, when you look through it at the light from the discharge lamp, you should see lines of colors "floating" beside the lamp.

> **Note:** If you have the grating oriented the wrong way, you won't see these lines – instead you'll see a rainbow effect above and below the lamp. Rotate your grating 90°.

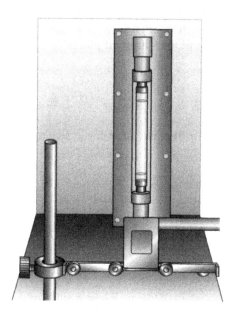

Figure 20-7: Placement of hydrogen discharge lamp.

3. Lay a meter stick in front of the lamp so that when you look through the grating, the colored lines appear to be floating above it. Measure and record in the table below the distance between the lamp and each of these lines.

	Line Color	Distance from Lamp
1		$x_1 =$ cm
2		$x_2 =$ cm
3		$x_3 =$ cm

4. Measure and record the distance between the lamp and the diffraction grating:

$$L = \underline{\hspace{2cm}} \text{ cm}$$

5. Your instructor will provide you with a different gas lamp. Repeat the set up for the new lamp, and observe the lines of color that you get from it. Describe 3 ways in which these lines are different from the ones you have recorded above.

Interpretation

1. Consider the question, "In this investigation, why do we see colored lines floating in space?" We know that the discharge lamp puts out light of different wavelengths (colors), mixed together. Each of the light waves, e.g., red, travels from the lamp straight to the diffraction grating, where it undergoes diffraction. Just as in Investigation 2, each wave will therefore cast a bright spot at some angle toward our eyes. Because the bright spot is at an angle, it appears to be coming from an angle away from the discharge lamp. So, explain why the different colors appear at different places?

2. For the first color line, calculate the angle at which it appears:

$$\theta_1 = \tan^{-1}\left(\frac{x_1}{L}\right) = \tan^{-1}\left(\frac{\underline{\hspace{1cm}} \text{ cm}}{\underline{\hspace{1cm}} \text{ cm}}\right) = \underline{\hspace{1cm}}^{\circ}$$

3. Using this angle, calculate the wavelength, λ_1, of the red line. Here, n is equal to 1 because we're only seeing the first bright red spot, out from the middle.

$$n\lambda_1 = d \sin\theta_1$$

$$\lambda_1 = d \sin\theta_1 = (\underline{\hspace{1cm}} m) \times \sin(\underline{\hspace{1cm}}^{\circ}) = \underline{\hspace{1cm}} m$$

(Use the d you calculated for the grating in Investigation 1)

4. Repeat step 3 and 4 for the other lines you recorded. Write your results in the table below.

	Line color	Angle θ	Wavelength λ
1			$\times 10^{-9}$ m
2			$\times 10^{-9}$ m
3			$\times 10^{-9}$ m

5. Look up these wavelengths in the color table given in the Background Concepts section of this experiment. Do the colors that the table gives for the wavelengths match the colors you saw for each line? How well do they match?

Report Sheet

Name: _____

Experiment 20

Date: _____ Section: _____

1. Using the results from Investigation 1, fill in the following data table:

	Angle of bright spots		Wavelength of laser
n = 1	θ_1 =		λ =
n = 2	θ_2 =		λ =

spacing of slits: d = _____

2. Using the results from Investigation 2, fill in the following data table:

Distance between sources (d)	
Angle of side rays (θ)	
Calculated wavelength ($\lambda_{calculated}$)	
Percent deviation from $\lambda_{measured}$	%

3. Using the results from Investigation 3, fill in the following table:

	Line color	Angle, θ	Wavelength, λ
1			$\times\ 10^{-9}$ m
2			$\times\ 10^{-9}$ m
3			$\times\ 10^{-9}$ m

4. Meticulous measurements of the hydrogen spectrum give the following values for the three most visible lines:

656 nm

486 nm

434 nm

How do these compare with the values you calculated in Investigation 3?

5. When gases of different elements are energized, each produces its own unique line spectrum. Now, let's say we have two gas lamps, each with a different mixture of gases. They produce the following line spectra:

Gas Lamp A	Gas Lamp B
710 nm	666 nm
668 nm	632 nm
656 nm	605 nm
589 nm	599 nm
502 nm	546 nm
486 nm	549 nm
469 nm	435 nm
434 nm	404 nm
405 nm	

Based on these values, is it likely that either gas lamp contains some hydrogen gas?

6. If you knew the "line spectrum" for each element, how could you figure out what elements are present in the Sun?

7. For a precise measurement of a line spectrum, we would use a very finely spaced grating (small d). If instead you used a diffraction grating with a larger d, all the θ's would be smaller. Why would that decrease the precision of your results?

Extra Credit

1. In Investigation 3, you might have noticed a second red line, a second violet line, or a second blue line, each more faint, appearing further out from the lamp. These are lines corresponding to the second order of diffraction (i.e., $n = 2$). Calculate at what angles these lines might appear, and the corresponding values of x.

How can you determine the focal length of a diverging lens?

Optics of Thin Lenses

■ Background Concepts

Lenses rely on refraction to redirect light rays in useful ways. Converging (convex) and diverging (concave) lenses are the two most basic types.

Converging lenses are generally convex on both sides. Because of their shape, all rays of light that comes in straight (parallel to the optic axis) are redirected toward the same point on the other side of the lens. The point where these rays converge is called the focal point. If you've ever used a magnifying glass, then you've used a converging lens.

Diverging lenses are generally concave on both sides. Because of their shape, all rays of light that are parallel to the optic axis of the lens are redirected outward. It is as if all of these rays had come from the same point in front of the lens and are diverging outward. The point in front of the lens that appears to be the origin of the rays is the focal point of the lens. When working with simple lenses it makes no difference which side is considered front or back. If a lens is not concave or convex on both sides, it may be either a converging or a diverging lens.

When light rays from a single object pass through a lens, they appear to have come from another location. An observer seeing the final light rays sees an image of the object at that location. In one case, the light rays actually come from the image, having been focused there by the lens. These are called *real* images, and can be made to appear on a screen. In the other case, the light rays do not actually pass through the image's location. The directions of these rays only make it appear that they come from or through this point. Since the light rays never really passed through this point, it is called a *virtual* image. Even if a screen is placed at that point, the image will not appear, because the light rays do not actually pass through.

In both instances, the location of the image can be calculated using the following equation:

$$\frac{1}{s} + \frac{1}{s'} = \frac{1}{f}$$

where s measures how far the object is in front of the lens, s' measures how far the image appears from the lens, and f is the *focal length* of the lens, that is, the distance from

---

the lens to the focal point. The values for the focal length are considered positive for converging lenses and negative for diverging lenses. The object distance, s, is almost always positive, since the object must necessarily be in front of the lens – except in certain instances. The image distance, s', is positive for real images that appear beyond the lens, and negative for virtual images that appear before the lens.

The image that a lens casts can be larger or smaller than the actual size of the object. An image that is twice as large as the object is said to have been magnified by a factor of 2. If it were twice as large, but upside down, we would say that its magnification is – 2. A magnification of + 0.5 means that the image is half the size of the object and right side up. The equation for calculation of magnification, m, is:

$$m = \frac{\text{image size}}{\text{object size}} = -\frac{\text{image distance}}{\text{object distance}} = -\frac{s'}{s}$$

In this equation, a negative sign is included to account for the *inverted* real image (negative m) formed behind the lens, by an object placed in front of the lens (both image and object distances positive).

When the light rays from an object pass through two lenses, the second lens "sees" the rays from the first lens as its object. In this situation, the image of the first lens is used as the object for the second lens, when calculating where the image of the second lens will form.

In this Experiment, you will be given three lenses: two converging and one diverging. Your challenge is to determine the focal lengths of each converging lens. You will then use your measurements for the converging lenses to determine the focal length of the diverging lens.

Investigation 1

Objectives

To project real images of an object using two separate converging lenses, calculate the focal length of each lens based on the characteristics of the images projected, and calculate the magnification of the images.

Procedure

> **NOTE:** If possible, dim the room lights for this investigation.

1. Your instructor will provide you with an optics bench, including a screen, a lantern, lens holders, and three lenses: 2 converging and 1 diverging. It will be your goal in Investigation 1 to determine the focal lengths of the two converging lenses.

2. Examine the three lenses as shown in Figure 21-1, and determine which one is the diverging lens. (The one that does not make distant objects appear upside down.)

Figure 21-1

3. Pick one of the converging lenses and call it Lens A. Attach it to a lens holder and place it on the optics bench.

4. As shown in Figure 21-2, place the lantern at the leftmost edge of the optics bench and turn it on. Note: your lantern might actually consist of two separate parts: a light source and a cut-out shade plate. If so, place the light source to the left of the optics bench and place the shade plate in front of it, so that the light shines through the shade plate, toward the lens. The lantern's shade plate will be used as the object.

5. Adjust the position of the lens so that it is about 30 cm from the object. Attach the screen to the track using the appropriate holder. Position it about 30 cm beyond the lens.

6. Vary the position of the screen so that the distance between it and the lens changes between 10 and 70 cm. Adjust the position of the screen to make the image as sharp as possible.

7. Measure and record the size of the image or some part of the image, such as the arrowhead.

$$h_1 = \underline{\hspace{2cm}} \text{ cm}$$

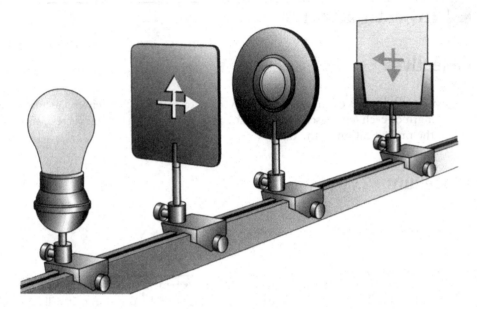

Figure 21-2: Position of lantern and shade plate.

8. Measure the size of the corresponding part of the object

$$h_o = \underline{\hspace{2cm}} \text{ cm}$$

9. Measure and record in Table 21-1 (Case #1) the distance, s, between the object and the lens, and the distance, s', between the lens and the image.

	Object distance, s	Image distance, s'	Focal length, f
Case #1	cm	cm	cm
Case #2	cm	cm	cm
Case #3	cm	cm	cm

Table 21–1: Lens A data.

10. Move the lens backwards or forwards by at least 5 cm and change the position of the screen so that the sharp image is again formed on the screen. Repeat the measurements from step 9 and record them in the Table 21-1 (Case #2).

11. Again, move the lens and screen positions by at least 5 cm and find yet a different set of distances at which a sharp image is again formed. Repeat the measurements from step 9 and record them in the Table 21-1 (Case #3).

12. Remove the lens and replace it with the other converging lens (Lens B). Repeat the measurements of s and s' for the three cases, using Lens B. Record your values in the Table 21-2. (Again, make sure that each set of measurements differs by at least 5 cm.)

	Object distance, s	Image distance, s'	Focal length, f
Case #1	cm	cm	cm
Case #2	cm	cm	cm
Case #3	cm	cm	cm

Table 21–2: Lens B data.

Interpretation

1. Using the data from Lens A (Case #1), calculate the following:

$$\frac{1}{s} = \underline{\hspace{2cm}}$$

$$\frac{1}{s'} = \underline{\hspace{2cm}}$$

$$\frac{1}{f} = \frac{1}{s} + \frac{1}{s'} = \underline{\hspace{2cm}}$$

$$f = \underline{\hspace{2cm}} \text{ cm}$$

   Record this value of f in Table 21-1 (Case #1).

2. In a similar way, calculate the value of f for each of the other cases in Tables 21-1 and Table 21-2.

   What is the average f for lens A? _____ cm

   What is the average f for lens B? _____ cm

3. In all of these cases, were the images upright or inverted, as compared to the object?

4. Using the image and object sizes measured in steps 7 and 8 of the Procedure, calculate the magnification:

$$m = \frac{h_i}{h_o} = \underline{\hspace{2cm}}$$

5. Calculate the ratio:

$$-\frac{s'}{s} = \underline{\hspace{2cm}}$$

   Does this equal the results obtained in question 4?

6. How does turning the lens so you're looking through the opposite side affect the image? (Try it).

Investigation 2

Objectives

To project an image through two converging lenses onto a screen, calculate the focal length of the second converging lens based on the location of the final image and compare this focal length to the value obtained in Investigation 1.

Procedure

1. As in Investigation 1, set up the converging lens with the longer focal length, lens$_l$, to project a sharp image onto the screen. Place the lens$_l$ about 40 cm away from the lantern. (If you are using a stand-alone lantern as your object, place the lens at the edge of the optics bench, and place the lantern about 40 cm away from the bench to conserve bench space.) Measure and record the object distance and image distance.

$$s_l = \underline{\hspace{2cm}} \text{ cm}$$

$$s_l' = \underline{\hspace{2cm}} \text{ cm}$$

2. Place the other converging lens$_s$ (the one with the shorter focal length) about 20 cm behind the screen. Measure and record the distance in front of lens$_s$ the screen image from the lens$_l$ is located. This is the object distance for lens$_s$.

$$s_s = \underline{\hspace{2cm}} \text{ cm}$$

3. Now remove the screen and place it behind lens$_s$. Adjust the screen's location until a sharp image is formed on it from lens$_s$. (Note: it may be necessary to change the position of lens$_s$ and the distance recorded in step 2). Measure and record the distance between lens$_s$ and the final image location on the screen. (It may be necessary to measure beyond the edge of the bench.)

$$s_s' = \underline{\hspace{2cm}} \text{ cm}$$

Interpretation

1. As in the interpretation of Investigation 1, calculate the focal length for lens$_s$ using the object distance and image distance from steps 2 and 3 above.

$$f_s = \underline{\hspace{2cm}} \text{ cm}$$

2. Calculate the percent deviation between the above value for f_s, and the average value calculated for the *same* lens in Investigation 1:

$$\text{percent deviation} = \frac{|\text{value Investigation 1} - \text{value Investigation 2}|}{\text{value Investigation 1}} \times 100 = \underline{\hspace{1cm}}\%$$

3. Is the final image inverted or upright compared to the original object?

4. Is the final image inverted or upright compared to the image formed by lens$_l$?

5. After you remove the screen from between the lenses in step 3, the image from lens$_l$ no longer falls on a screen. At this point, is the image from lens$_l$ real or virtual?

Investigation 3

Objectives

To project an image onto a screen using a combination of one converging and one diverging lens and calculate the focal length of the diverging lens based on the location of the final image.

Procedure

1. As you did in Investigation 1, project an image onto a screen using one of the converging lenses. Use converging lens$_S$, the one that has the smaller focal length. Adjust the lens and screen positions so that a sharp image is achieved about 40 cm from the lens. Record the object distance and image distances.

$$s_S = \text{_____} \text{ cm}$$

$$s_S{}' = \text{_____} \text{ cm}$$

2. Place Lens C (the diverging lens) about 10 cm in *front* of the screen, without changing the position of the screen. Measure and record the distance between the position of Lens C and the screen, where the sharp image was before you put in Lens C. This is the object distance for Lens C.

$$s_C = -\text{_____} \text{ cm}$$

(Note the distance for a diverging lens is negative.)

3. Now remove the screen and place it further back, where Lens C projects a sharp image onto it. Measure and record the distance between lens C and the final sharp image.

$$s_C{}' = \text{_____} \text{ cm}$$

Interpretation

1. As you did in Investigation 1, calculate the focal length of Lens C using the object distance and image distances determined in steps 2 and 3 above.

$$f_C = \text{_____} \text{ cm}$$

(Note it should be negative.)

2. Is the final image upright or inverted as compared to the object?

3. Is the image from the first lens upright or inverted as compared to the object?

4. Why is the object distance, s_C, negative in this case?

Report Sheet

Name: _____

Experiment 21

Date: _____ Section: _____

---

1. Present your best values for the focal lengths of each of the three lenses.

$$f_A = \text{_____} \text{ cm}$$

$$f_B = \text{_____} \text{ cm}$$

$$f_C = \text{_____} \text{ cm}$$

2. Record the percent deviation from Investigation 2.

_____ %

3. Presumably, the percent deviation in our results for f_B speaks to the uncertainty inherent in our experimental method for determining focal lengths. Assume that the same percent deviation applies to the result obtained for f_C. Given this level of uncertainty, within what range of values can we expect the "true" value of f_C to lie?

$$f_{C \text{ minimum}} = \text{_____} \text{ cm}$$

$$f_{C \text{ maximum}} = \text{_____} \text{ cm}$$

4. In the space below, sketch diagrams of the setup from Investigations 2 and 3, showing the placement of the object, lenses, first image, and final image.

Investigation 2

Investigation 3

5. Why can't a diverging lens be used to project an image in Investigation 1?

6. The power of a lens is commonly rated in *diopters*, where power in diopters = $1/f$, and f is expressed in cm. Using this definition, express the optical power of your three lenses in diopters.

7. If you have a series of five converging lenses, how can you calculate where the final image will appear?

Extra Credit

1. A camera with a single lens is to be used to take a picture of a person. The person is 1.8 meters tall, and is standing 3 meters in front of the camera. It is desired to have the image of the person fill the entire height of the 35 mm film, placed 5 cm behind the lens. What focal length does the camera's converging lens need to have to accomplish this?

2. A converging lens with a focal length of 15 cm is used to project an image of the Sun onto a screen. The Sun is so far away that we can approximate the object distance as infinite. Where, then, does the sharp image appear behind the lens?

Why are some solutions colored?

Spectrophotometry

Background Concepts

Electromagnetic radiation contains many different wavelengths. As it passes through a substance, some wavelengths are absorbed and the rest are either transmitted or reflected. The various regions of the electromagnetic spectrum are composed of photons with differing amounts of energy. The amount of energy can be calculated using the equation $E = hc/\lambda$, where h is Planck's constant (6.626×10^{-34} J/Hz), c is the speed of light (3.00×10^8 m/s), and λ is the wavelength (color) of light. Experimental results indicate that X ray photons have approximately the same amount of energy as the forces holding the inner, non-bonding electrons to the nucleus. Photons in the ultraviolet and visible regions of the spectrum have approximately the same amount of energy as electrons in chemical bonds. The energy of infrared photons matches the energy of the vibrational, stretching, and twisting motions of molecular bonds. It is, therefore, the electron and chemical bond arrangements in a substance that determine if it will ab-

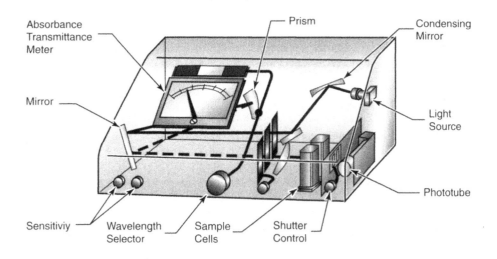

Figure 22-1: Spectrophotometer

sorb a particular wavelength or not. This, in turn, is responsible for the color of the solution we see. Therefore, light can be used to help identify both the kind and amount of a substance present in a solution.

The spectrometer is an instrument that measures the intensity of various wavelengths of light. As shown in Figure 22-1, all spectrometers have the same basic components. Its essential features are a light source, a series of mirrors and lenses, a prism or grating to select the desired wavelength, a place to insert the sample, a photodetector, and a meter. In spectrophotometers operating in the visible light range, the light source may be a sophisticated version of the tungsten light bulb used at home. The light from this source is passed through a prism, or grating that spreads it into its individual colors so that a particular wavelength can be passed through the sample.

Since we know that chemical bonds can absorb light in the visible region, it is possible to identify a compound by determining what wavelengths are absorbed. For example, a substance having two double bonds will absorb about twice as much energy as one having only one double bond. The intensity of the absorption, therefore, can be used to determine both the number and kinds of chemical bonds involved.

Of the wavelengths absorbed, the one that has the greatest absorbance is called "lambda max," λ_{max}. As it suggests, λ_{max} is the wavelength that is at the center of the absorbance peak. It is also the one often used to detect the presence of that substance in a sample. The intensity of this absorption is called the compound's absorptivity. When examining the amount of light passing through a sample, you consider either the amount of light that passes through, called *transmittance*, or the amount of light absorbed. Transmittance, T, is read from the spectrophotometer as a percentage. A transparent material has 100% transmittance and an opaque material has 0% transmittance. Absorbance, A, is the measure of the amount of light that does not pass through the sample. The mathematical relationship between absorbance and percent transmittance is the following:

$$\text{Absorbance} = 2 - \log \% \, T$$

In this experiment, we will use light from the visible portion of the spectrum (400 – 800 nm) to determine λ_{max} for a test solution and compare its absorbance peak to its color.

Investigation 1

Objective

To determine the absorbance peak and λ_{max} for a colored solution.

Safety Requirements

■ Always wear splash proof safety goggles.

■ The compounds used for this investigation may be mildly toxic and may produce significant stains on skin and clothes·

■ Use care when transferring solutions. Wash skin that is exposed to chemicals in cold water, then notify the teacher.

■ Gloves and a laboratory apron or lab coat are recommended·

■ Discard the waste chemicals as instructed by your teacher.

Procedure

1. Follow your instructor's directions for the correct operation of the spectrophotometer. For example, to assure reproducible results, most spectrophotometers need a 15 – 20 minute warm up period before calibration.

2. Obtain a pair of clean cuvettes. Examine them for cracks and scratches and note the orientation mark for proper insertion into the spectrophotometer.

3. Your instructor will assign you a colored test solution and indicate the recommended level that should be put into the cuvette each time to obtain an accurate reading.

4. Once the spectrophotometer is properly warmed, close the empty sample chamber and with the light beam blocked, adjust the "zero transmittance" knob until the meter reads zero transmittance.

5. Adjust the wavelength setting of your spectrometer to approximately the middle of the range you will be scanning. In this case, it would be 600 nm.

6. Fill both cuvettes to the recommended level with DI water.

7. While holding the cuvette by its top, use a lab tissue to wipe any excess water, smudges, or fingerprints from its outside.

8. Insert the cuvettes of DI water into the spectrophotometer, making sure that it is properly oriented.

9. Using the % T, adjust knob to set the spectrophotometer to read 100% transmittance.

10. Remove the first cuvette and place the second cuvette of DI water into the spectrophotometer, making sure that it is properly oriented.

Test Solution Number:			Test Solution Number:		
Data Source:			Data Source:		
Test Solution Color:			Test Solution Color:		
Standard Cuvette % T:			Standard Cuvette % T:		
λ	% T	Absorbance (Calculated)	λ	% T	Absorbance (Calculated)
800			800		
780			780		
760			760		
740			740		
720			720		
700			700		
680			680		
660			660		
640			640		
620			620		
600			600		
580			580		
560			560		
540			540		
520			520		
500			500		
480			480		
460			460		
440			440		
420			420		
400			400		

Table 22–1: Percent transmittance data for the two test solutions.

11. (Option 1) If this cuvette of DI water gives a reading *less* than 100% T, then it will be your "standard" cuvette. Record in Table 22-1 its % T reading. Remove the cuvette and keep it. At each new wavelength setting, reinsert this cuvette and re-adjust the % T to this value. (The first cuvette should be emptied and used for your test solution.)

    (Option 2) If this cuvette of DI water gives a reading *more* than 100% T, then readjust the transmittance so it again reads 100% T. Remove the cuvette and put the first cuvette of DI water back into the spectrophotometer. Read and record in Table 22-1 the % T. Remove this cuvette and keep it as the "standard" cuvette. At each new wavelength setting, reinsert this cuvette and adjust the % T to this value. (The other cuvette should be emptied and used for your test solution.)

12. Adjust the wavelength of the spectrometer to 800 nm. Insert the "standard" cu-vette and adjust the % T to the same value as you recorded in step 11.

13. Use a laboratory tissue to dry the inside of the test sample cuvette, then fill it to the designated level with the test solution assigned.

14. Wipe any excess test solution, fingerprints, or smudges from the outside of the cuvette. Insert the sample cuvette into the spectrophotometer, again being care-ful to orient it properly.

15. Record the spectrophotometer reading for % transmittance in Table 22-1.

16. Remove and save the cuvette and its contents.

17. Adjust the spectrophotometer wavelength setting to 20 nm less than the previous setting.

18. Reinsert the "standard" cuvette and readjust the instrument to the same % T reading as recorded in step 11. Remove the "standard" cuvette and save it.

19. Reinsert the sample cuvette and read and record the % T in Table 22-1.

20. Repeat steps 16 – 19 until data has been collected for each of the wavelengths be-tween 800 and 400 nm.

Interpretation

1. Using the % T data, your calculator, and the following formula, complete the Absorbance column in Table 22-1 for your test solution.

$$\text{Absorbance} = 2 - \log \% \, T$$

2. Consider the range of the absorbance data in Table 22-1 and select an appropriate scale to represent it on the *y*-axis. Prepare Graph 22-1 of Absorbance vs. Wave-length. Use a pencil to lightly draw a smooth line among the points.

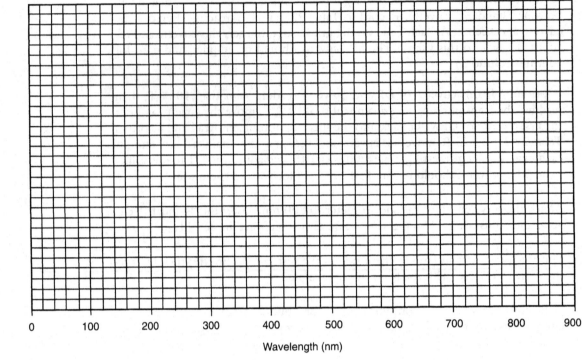

Graph 22–1: Absorbance vs. wavelength.

3. Review the graph. Is there a region in which more data would help define the shape of the curve? Identify five additional wavelengths to measure, each separated by at least 5 nm. Record the wavelengths you have chosen at the bottom of Table 22-1.

4. Using the same standard and test sample cuvettes, measure the % T for each of these wavelengths and record your data in Table 22-1.

5. Perform the necessary calculations so that you can add this data to your graph. Redraw the smooth line, if necessary, for the graph.

6. Does your test solution have more than one absorbance peak?

7. What is the width of the major (if more than one) absorbance peak? (answer in nm)

8. Using the absorbance peak wavelength and the chart below, determine the color at which your test sample absorbed most significantly.

Red	Orange	Yellow	Green	Blue	Violet
630 – 770 nm	590 – 630 nm	560 – 590 nm	490 – 560 nm	430 – 490 nm	380 – 440 nm

9. How does the color of your test sample compare to the color at which your test sample absorbed most significantly? Explain.

10. Based on your absorbance peak, what is "lambda max," λ_{max}, for your test solution?

Investigation 2

Objective

To determine lambda max, λ_{max}, for another test solution.

Procedure

1. Identify a student who has one of the other test solutions.

2. Exchange your % T and color data and add it to Table 22-1.

3. Complete all the calculations necessary to complete your Table 22-1.

Interpretation

1. Add the data for the other test solution to Graphs 22-1. Be sure to identify each curve.

2. What are the similarities and differences between the test solution curves?

3. Does the new test solution have more than one absorbance peak?

4. How broad is its major (if more than one) absorbance peak? (answer in nm)

5. How does the color of the test sample compare to the color at which the test sample absorbed most significantly? Explain.

6. Based on the absorbance peak, what is "lambda max," λ_{max}, for this test solution?

    Test Solution No. _____: λ_{max} = _____

Investigation 3

Objective

To examine the relationship between λ_{max} and a solution's color.

Procedure

1. Figure 22-2 is a circle that has been divided into six equal segments.

2. Prepare a Complementary Color Wheel by starting at any one of the segments and proceeding clockwise around the wheel writing, in order, the following color names – red, orange yellow, green, blue, and violet.

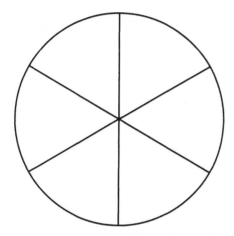

Figure 22-2: Complementary Color Wheel

Interpretation

1. Using the chart presented in Investigation I, complete the following for each of the test solutions:

        Test Solution No. _____:

           λ_{max} = _____ = Absorbed color _____

        Test Solution No. _____:

           λ_{max} = _____ = Absorbed color _____

2. Locate the absorbed color and the solution's color on the Complementary Color Wheel for each of the test solutions. Is there a pattern between the absorbed color and the test solution's color? Explain your findings.

Report Sheet

Name: _____

Experiment 22

Date: _____ **Section:** _____

1. Complete the following table by transferring your % T and absorbance data from Table 22-1

Test Solution Number:			Test Solution Number:		
Data Source:			Data Source:		
Test Solution Color:			Test Solution Color:		
Standard Cuvette % T:			Standard Cuvette % T:		
λ	% T	Absorbance (Calculated)	λ	% T	Absorbance (Calculated)
800			800		
780			780		
760			760		
740			740		
720			720		
700			700		
680			680		
660			660		
640			640		
620			620		
600			600		
580			580		
560			560		
540			540		
520			520		
500			500		
480			480		
460			460		
440			440		
420			420		
400			400		

2. Attach Graph 22-1 to your Report Sheet.

3. Complete the following with λ_{max} and its absorbed color for each test solution.

        Test Solution No. _____:

                    λ_{max} = _____ = absorbed color _____

        Test Solution No. _____:

                    λ_{max} = _____ = absorbed color _____

4. When you compare the color absorbed to the solution's color for each of the test solutions, what relationship did you find? How do you explain this finding?

Extra Credit

1. The % T of a test solution was measured over a range of wavelengths and converted into its equivalent absorbance. The following graph of Absorbance vs. Wavelength was prepared. Examine Graph 22-2 and answer the following questions:

    a. What is the width of the absorption peak?

    b. What is λ_{max} for this test solution?

    c. What color is the solution?

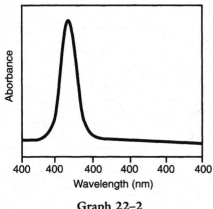

Graph 22–2

Does the arrangement of atoms within a molecule make a difference?

Molecular Models

Background Concepts

Chemical formulas are representations of the number and kind of atoms present in one molecule of a pure substance. Within the area of organic chemistry, three different types of formulas are used: molecular, full structural, and condensed structural.

C_2H_6O	**Molecular Formulas** – indicate only the number and kind of atoms in a molecule.
H H H–C–C–O–H H H	**Full Structural Formulas** – not only indicate the number and kind of atoms per molecule, but also show the sequence and the types of bond(s) connecting each atom.
CH_3–CH_2–OH CH_3CH_2–OH	**Condensed Structural Formulas** – indicate the same information as the full structural formula, but take up less space because single bonds are often omitted.
CH_3–CH_2–OH CH_3–O– CH_3	**Isomers** – When two or more chemical compounds have the same molecular formula, but differ in their structural formulas they are said to be isomers. As a result of the structural differences, isomers have different chemical and physical properties.

The major type of bonding in organic compounds is covalent bonding. Depending on the number of electron pairs being shared, covalent bonds are subdivided into three different types:

Type of Bond	No. of Electron Pairs Shared	Represented by
Single Covalent	1	A stick or spring
Double Covalent	2	Two sticks or springs
Triple Covalent	3	Three sticks or springs

The number of covalent bonds an element can form is called its covalence number. In model sets, atoms representing various elements are typically provided with the appropriate number of "holes" or "connectors" to satisfy that element's covalence number, as is shown in the commercial stick-and-ball type model illustrated at right.

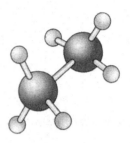

Element	Symbol	Color	Covalence Number
Carbon	C	Black	4
Hydrogen	H	Yellow	1
Oxygen	O	Red	2
Chlorine	Cl	Green	1

Investigation 1

Objective

To determine the type of carbon bonding and general formula for members of the alkane, alkene, and alkyne families.

Procedure

1. Select two carbon atoms, six hydrogen atoms, and the necessary connectors. Assemble the alkane structure and record its molecular, full structural, and condensed structural formulas below.

Molecular Structural Formula	Full Structural Formula	Condensed Structural Formula

   a. Grasp the two carbon atoms and attempt to rotate one with respect to the other. Will they rotate freely?

   b. What type of covalent bond connects the two carbon atoms?

2. Select two carbon atoms, four hydrogen atoms, and the necessary connectors. Assemble the alkene structure and record its molecular, full structural, and condensed structural formulas below.

Molecular Structural Formula	Full Structural Formula	Condensed Structural Formula

   a. Will the two carbon atoms in this model rotate freely?

   b. What type of covalent bond connects the carbon atoms?

3. Select two carbon atoms, two hydrogen atoms, and the necessary connectors. Assemble the alkyne structure and record its molecular, full structural and condensed structural formulas.

Molecular Structural Formula	Full Structural Formula	Condensed Structural Formula

a. Will the two carbon atoms in this model rotate freely?

b. What type of covalent bond connects the carbon atoms?

Interpretation

The general formulas C_nH_{2n+2}; C_nH_{2n}; and C_nH_{2n-2} can be used to represent the various hydrocarbon families. In the spaces below, write the formula for the compounds prepared in steps 1, 2, and 3 above, the type of covalent bonding between the two carbon atoms, and the general formula that represents the organic family.

Organic Family	Formula of Compound	Covalent Bond Type	General Formula
Alkane			
Alkene			
Alkyne			

Investigation 2

Objective

To determine if isomers exist for a particular molecular formula.

Procedure

1. Construct a possible model for the compound that has the formula C_3H_8O.

Condensed Structural Formula	

Using the same number of atoms and connectors, attempt to construct two other models that are isomers of the previous compound.

Condensed Structural Formula	
Condensed Structural Formula	

2. Construct a possible model for the compound that has the formula C_2H_5Cl.

Condensed Structural Formula	

   a. Do isomers exist for this compound?

3. Using the model constructed for step 2, replace one of the remaining hydrogen atoms with a second chlorine atom. Does it make a difference which of the remaining hydrogen atoms is replaced?

    a. Draw structural formulas for any differences that you found.

Structural Formula	

    b. Are these compounds isomers?

Interpretation:

1. Consider the formulas for the compounds prepared in Investigation 1, steps 1, 2, and 3. Were any or all of these compounds isomers? Explain your answer:

2. Consider the structural formulas for the compounds prepared in steps 1, 2, and 3 above. What determines if an isomer can exist? Explain:

Investigation 3

Objective

To determine the isotopes that exist for the compound C_4H_9Cl.

Procedure

1. Prepare all the possible isomers you can find for C_4H_{10}. Draw condensed structural formulas for each of the isomers you find.

Isomers	Condensed Structural Formulas

2. Select one of the isomers for C_4H_{10} you found in step 1 and replace one of the hydrogen atoms with a chlorine atom. How many places can you find to put the chlorine atom that makes a difference without changing the arrangement of the carbon atoms? Write condensed structural formulas for each of the isomers you found.

Isomers	Condensed Structural Formulas

3. Select the other isomer for C_4H_{10} you found in step 1 and replace one of the hydrogen atoms with a chlorine atom. How many different places can you find to put the chlorine atom without changing the arrangement of the carbon atoms? Write condensed structural formulas for each of the different isomers you found.

Isomers	Condensed Structural Formulas

Interpretation

1. What is the total number of isomers found for the molecular formula C_4H_9Cl?

_____ isomers

2. Write a brief explanation for why you consider these to all be isomers.

3. Write the condensed structural formulas for each isomer in the space below and provide a correct IUPAC name for each structure.

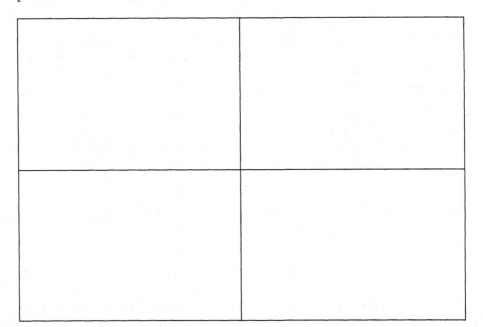

Investigation 4

Objective

To determine all the isotopes for the molecular formula $C_4H_{10}O$.

Procedure

1. Using the technique developed in Investigation 3, attempt to prepare all of the possible isomers for $C_4H_{10}O$. Draw condensed structural formulas for each of the isomers you find.

Condensed Structural Formulas

Interpretation

1. Select another group or person and compare the structural formulas you have found. Compare the position of each of the atoms to make sure they are not the same. Write the structural formulas for any additional isomers you have found.

Condensed Structural Formulas

2. Members of the alcohol family have the general formula of R–OH and members of the ether family are R–O–R. Subdivide all the isomers you have found and rewrite them under the appropriate headings.

Alcohols, R–OH	Ethers, R–O–R

Report Sheet

Name: _____

Experiment 23

Date: _____ Section: _____

1. Complete the following table using the information that you developed in Investigation 1.

Family	Formula of Compound	Covalent Bond Type	General Formula
Alkane			
Alkene			
Alkyne			

2. List the isomers you found in Investigation 2 for C_3H_8O.

3. Write the condensed structural formulas and IUPAC names for each of the isomers of C_4H_9Cl you found in Investigation 3.

4. List the isomers of $C_4H_{10}O$ you found, grouped under the appropriate family heading.

Alcohols, R–OH	Ethers, R–O–R

Extra Credit

Select one isomer from each family list in question 4. Using your textbook, a reference book, or the Internet compare their melting and boiling points, density, water solubility, chemical properties and IUPAC names.

What do artificial flavorings, aspirin, and soft drink bottles have in common?

Organic Esters

Background Concepts

Organic esters are compounds produced by catalytic removal of water (–OH and –H) from between the organic acid and alcohol functional groups. In nature, the reaction is catalyzed by enzymes but in the laboratory, sulfuric (H_2SO_4) or phosphoric (H_3PO_4) acids are typically used. The general equation for the reaction is:

$$\underset{\text{Organic acid}}{R-\overset{\displaystyle O}{\overset{\|}{C}}-O-H} \quad + \quad \underset{\text{Alcohol}}{H-O-R'} \quad \xrightarrow{H_2SO_4} \quad \underset{\text{Ester}}{R-\overset{\displaystyle O}{\overset{\|}{C}}-O-R'} \;+\; H_2O$$

The new bond formed connecting the remaining fragments is called an ester bond or ester linkage. It is both the place where the molecules joined and the place where acid and basic hydrolysis will later break the ester molecule apart.

Naturally Occurring Esters

The characteristic odors of many fruits and flowers are due to the presence of smaller naturally occurring esters. Commercially, these esters are synthetically produced and used in the production of both perfumes and artificial flavorings. Some of the common esters used as artificial flavorings and their associated aromas are:

Common Esters	Associated Aroma	Common Esters	Associated Aroma
Methyl butyrate	Apple	Octyl acetate	Orange
Pentyl propionate	Apricot	Isobutyl formate	Raspberry
Methyl anthranilate	Grape	Methyl salicylate	Wintergreen

Some esters also have medicinal value. Two of these esters are oil of wintergreen, which is used in analgesic rubs for sore muscles, and aspirin, which is used in many pain relieving and fever reducing pharmaceuticals.

Mother Nature also makes many larger, odorless esters commonly referred to as fats and waxes. If the ester is composed of three fatty acid molecules connected by the trihydric alcohol called glycerol, then it is called a triglyceride. Triglycerides are present in both animal fats and vegetable oils. The waxes used to polish cars (carnauba wax) and shoes (beeswax) are of this type. Fruits such as apples are coated by a thin layer of wax of this type. The reason you can polish an apple for your teacher, therefore, is due to this natural wax coating.

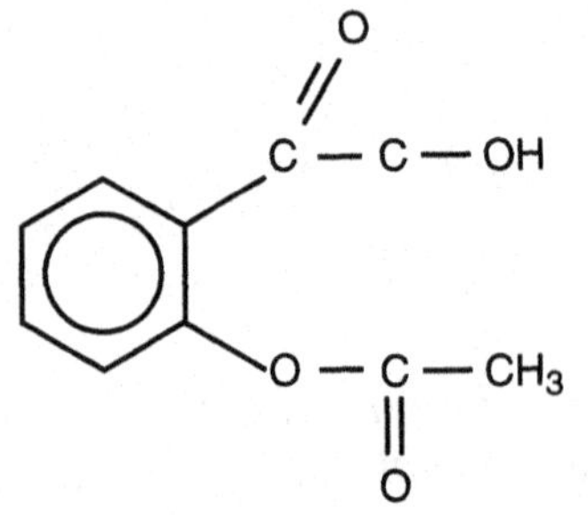

Figure 24–1: Acetyl salicylic acid.

Polyesters

Within the last 100 years, chemists discovered how to synthesize polyesters. It was found that if a dicarboxylic acid (having two acid functional groups) and a dihydric alcohol (having two alcohol functional groups) are reacted, the reaction will result in a polyester. That is to say, if molecule A bonds to molecule B, both molecules still have reactive functional groups available for reacting with the other type of molecule. The result of this chain-like reaction becomes a repeating pattern of –A–B–A–B–A– type, with an ester linkage between each A–B pair. The very large molecules resulting from this polymerization are known as polyesters. Careful examination of the tag on your clothing will likely reveal that it is some kind of cotton-polyester blend. Today, permanent press fabrics are very popular and comprise a large percent of our wardrobes.

Even more recently, there has been a shift away from using heavy glass containers to lighter polyester containers. Although we generally refer to them as "plastic" soda bottles, most are actually polyester containers. By examining the recycling insignia on its bottom, if you find PET or PETE under the recycling triangle that encloses a "1," then the bottle is composed of the ester polyethylene terephthalate. This important polyester is also known as Dacron® when it is used to make clothing fibers, and Mylar® when it is used to make videotapes, audiotapes, and bulletproof vests.

When you complete this experiment, you will be able to make some simple, useful esters. You will also be familiar with some of the characteristics of a common polyester. Finally, you may be able to answer the question "What do artificial flavorings, aspirin, and soft drink bottles have in common?"

Investigation 1

Objective

The objective of this investigation is to make several artificial flavorings.

Safety Requirements

- Wear eye protection at all times when in the laboratory.

- Use a test tube clip to hold or move hot test tubes.

- While heating a test tube, do not point it toward anyone or yourself.

- Avoid contact with all chemicals, especially sulfuric acid and acetic anhydride. If skin contact occurs with any chemical, flood the area with water and then notify the teacher.

- Use a clean or provided dropper to dispense reagents.

- Do not contaminate the reagent supply bottles by returning unused or unwanted reagents.

- Keep all chemicals well away from flames, unless intentional.

- Dispose of all chemicals according to the directions of your instructor.

Procedure

1. Prepare a hot water bath by placing about 25 mL of water in a 50 mL beaker. Use either a hot plate or a ring stand and burner (Figure 24-2) to heat the water to near boiling (80 – 100°C).

2. Label three 10 mL test tubes with A, B, and C.

3. In tube A, place 20 drops of glacial acetic acid and 20 drops of isoamyl alcohol. Cautiously note the odor of each and record in Table 24-1.

4. Using a clean stirring rod, mix the contents of the tube and then add 2 drops of concentrated H_2SO_4 *(caution)*. Place the test tube and its contents in the hot water bath and allow to boil for 3 minutes.

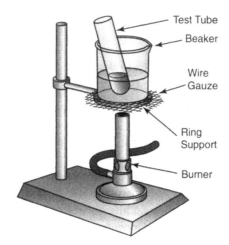

Figure 24–2: Hot water bath.

Tube	Carboxylic Acid Used	
	Formula	Odor
A		
B		
C		

Tube	Alcohol Used	
	Formula	Odor
A		
B		
C		

Tube	Ester Produced	
	Formula	Odor
A		
B		
C		

Table 24–1: Observation data.

5. Remove the tube from the hot water bath and set it aside to cool.

6. Note the odor and record it in Table 24-1.

7. Repeat the above procedure using the B labeled tube. Using the spatula provided or a clean spatula, transfer about 0.2 g (approximately the size of a green pea) of salicylic acid into the tube. Add 20 drops of methyl alcohol. Mix before adding 2 drops of concentrated H_2SO_4 *(caution)* and heating.

8. Record your observations for tube B in Table 24-1.

9. Repeat the above procedure using the C labeled tube. Place 12 drops of butyric acid and 20 drops of ethyl alcohol into the tube. Mix before adding 2 drops of concentrated H_2SO_4 *(caution)* and heating.

10. Record your observations for tube C in Table 24-1.

Interpretation

1. Describe the odor change, if any, that occurred after heating each of the carboxylic acid and alcohol mixtures.

2. What is the role performed by the H_2SO_4?

3. Complete the following table by matching each artificial flavor with the name of its ester.

Artificial Flavoring	Ester Name
Banana	
Pineapple	
Wintergreen	

4. List all the products you can think of that use methyl salicylate as an active ingredient or artificial flavoring agent.

5. *Never taste laboratory mixtures!* Would you expect the mixtures prepared in this Investigation to have tastes that resemble their odors? Why or why not?

Investigation 2

Objective

The purpose of this investigation is to prepare aspirin.

Procedure

1. Place about 0.5 g (approximately the size of 5 green peas) of salicylic acid in a 20 mL test tube.

2. Add 30 drops of acetic anhydride *(caution)* to the test tube. Note the odor of acetic anhydride *(caution)* and record below.

3. Using a clean stirring rod, stir the mixture and add 1 drop of concentrated H_2SO_4 *(caution)*.

4. Place the test tube in a hot water bath and boil for 3 – 5 minutes.

5. While still hot, cautiously add 10 drops of DI water to the test tube to decompose any excess acetic anhydride. Hot vapors of acetic acid will be given off.

6. Remove the test tube from the hot water bath and allow the tube to cool to room temperature. If crystals do not appear, place the tube in an ice bath and further cool the mixture. If crystals still fail to appear, use a clean stirring rod to "seed" your tube by dipping it into a tube that does have crystals and then back into your tube. After crystals appear, add 40 – 50 drops of DI water.

7. Set up a filter cone in a funnel as shown in Figure 24-3. Place a small beaker below the funnel stem to collect the filtrate. Transfer the contents of the test tube into the filter cone.

8. Wash the aspirin crystals on the filter paper using two 5 mL portions of cold DI water. The filtrate and rinse water washing can be discarded down the drain. Note the odor and color of the aspirin collected and record below.

9. Remove the filter paper, and using a clean spatula remove a small amount of the aspirin and place it in a clean 10 mL test tube.

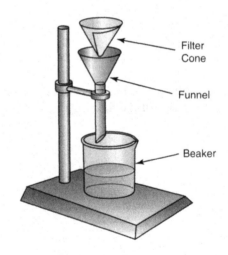

Figure 24–3: Funnel with filter cone.

10. Crush a commercially prepared aspirin tablet into a powder and place about the same amount of it in a second clean 10 mL test tube.

11. To each of the test tubes add 20 drops of DI water and then add 1 drop of ferric chloride, $FeCl_3$, test solution.

12. Note the color of each solution and record your findings below.

Interpretation

1. Describe the odor of acetic anhydride. Does the odor remind you of anything familiar?

2. Write the equation for the formation of aspirin, starting with salicylic acid and acetic anhydride.

3. Describe the odor and appearance of the aspirin produced.

4. What was the color of your aspirin solution after adding the $FeCl_3$ solution?

5. What was the color of the commercial aspirin solution after adding the $FeCl_3$ solution?

6. When $FeCl_3$ test solution contacts a phenol group, a violet color develops. When the aspirin you prepared is compared to the commercial aspirin, which had more unreacted phenol groups present?

7. If unreacted phenol groups mean less pure aspirin, which aspirin is more pure, yours or the commercially prepared? Explain your reasoning.

Investigation 3

Objective

The purpose of this investigation is to describe some of the properties of PET plastic.

Procedure

1. Obtain a small strip of polyethylene terephthalate (PET) plastic from the supply area. Using scissors, cut the strip into four 5 mm squares.

2. Place one square of PET into each of three 10 mL test tubes.

3. To the first test tube, add 10 drops of DI water and 2 drops of concentrated sulfuric acid (H_2SO_4) *(caution)*. Use a clean dry stirring rod to determine if PET will dissolve in a strong acid solution. Record your findings below.

4. To the second test tube, add 10 drops of DI water and 2 drops of 6 M sodium hydroxide (NaOH) solution. Use a clean dry stirring rod to determine if PET will dissolve in strong base solution. Record your findings below.

5. To the third test tube, add 10 drops of acetone. Use a clean dry stirring rod to determine if PET will dissolve in this organic solvent. Record your findings below.

6. Hold the fourth piece of PET with your forceps. Slowly pass the piece through a small burner flame several times. Describe its behavior below.

7. After the brief flame encounter, heat the PET directly in the flame. Describe its behavior below.

Interpretation

1. Describe your observations for PET and concentrated sulfuric acid.

2. Does it appear that PET could be used as a container for strong acids? Why or why not?

3. Describe your observations for PET and 6 M NaOH.

4. Does it appear that PET could be used as a container for strong bases? Why or why not?

5. Describe your observations for PET and acetone.

6. Does it appear that PET could be used as a container for organic solvents? Why or why not?

7. Describe your observations for the brief contact of PET with a flame.

8. Does it appear that PET could be used for applications where direct flame contact occurs? Why or why not?

9. Describe your observations for prolonged direct contact of PET with a flame.

10. Does it appear that PET could be incinerated as a fuel source in an energy recovery system? Why or why not?

Report Sheet

Name: _____

Experiment 24

Date: _____ Section: _____

1. Complete the table using the information gathered in Investigation 1. Write the formulas and names for the reactants and esters, corresponding to the specific artificial flavoring.

Artificial Flavoring	Acid Formula	Alcohol Formula	Ester Formula
Banana			
Pineapple			
Wintergreen			

2. Write a balanced equation for the formation of aspirin from salicylic acid and acetic anhydride.

3. Which aspirin was more pure, yours or the commercially prepared? Explain the basis for your answer.

4. Based on the tests performed, describe some of PET's physical and chemical characteristics.

5. What do artificial flavorings, aspirin, and PET soft drink bottles have in common? Explain your answer.

Extra Credit

PET is a polyester. Does that mean that polyesters are always PET? Why or why not? Explain your answer.

How do the properties of organic families differ?

Using Properties to Identify Organic Families

Background Concepts

Even before 1828, when Friedrich Wöhler made the first organic compound (urea) without the aid of a plant or animal, chemists recognized that organic compounds could be grouped based on their chemical and physical properties. Today there are 13 generally recognized organic families. Their names and general structural formulas are shown in Table 25-1.

Although there are small variations in the properties within each family, in general their behaviors are similar enough that a set of characteristics can be stated. For example, members of the alcohol family have a polar functional group that increase their water solubility, melting and boiling points. As in most organic families, as the length of the carbon chain becomes longer, their degree of water solubility diminishes. For example, methyl alcohol, CH_3–OH is completely soluble in water, while dodecyl alcohol, $C_{12}H_{25}$–OH, is not. Smaller family members ($C_1 - C_4$) of many organic families have some water solubility, but rapidly loose it as the length of the carbon chain increases. The degree of water solubility, therefore, becomes an important tool when identifying organic families.

In a similar way, organic compounds are either saturated or unsaturated and that can be easily determined by their reaction with a Br_2 test solution. In those organic families that do not have multiple carbon to carbon bonds, the orange/yellow color of the Br_2 remains, while those that have double or triple bonds react with the Br_2 forming a colorless solution. Another distinction between saturated and unsaturated compounds is the color of their flames when burning. Saturated hydrocarbons always burn with a blue flame, like a gas stove. If the organic family contains double or triple bonds, however, the flame is yellow and often produces black smoke. You may have noticed that a burning car tire produces bright yellow flames and lots of black smoke.

Some organic compounds have functional groups that interact with water resulting in either acidic or basic solutions. The carboxyl group, –COOH, that is a part of all carboxylic acids, ionizes in water producing the carboxylate ion, –COO$^-$, and a hy-

---

dronium ion, H_3O^+. Aqueous solutions of this family will therefore cause litmus to change from blue to red. In a similar way, amines $R-NH_2$, if they are soluble, gain a H^+ from the water and form an alkyl ammonium ion $R-NH_3^+$ ion and a hydroxide ion, OH^- ion. Aqueous solutions containing members of this family are therefore basic and change red litmus to blue.

When you look at the bigger picture, it becomes apparent that each organic family has a unique set of physical and chemical properties. By using a combination of these properties, it is possible to devise a scheme that will allow for family identification. Such a separation scheme is commonly referred to as a flow diagram. Simply stated, a flow diagram is a decision tree where the various branches are selected based on YES or NO answers. As a compound is subjected to the various tests called for in the flow diagram, it confirms or eliminates other choices until its family membership is narrowed to a single family. Once that family is identified, a confirmation test is typically performed to further insure its placement.

In this experiment, you will start with representative compounds from five different organic families. By using their solubility and reactions with other chemicals, you will develop a flow diagram that will allow you to identify the organic family. Once developed, you will be asked to correctly identify the family membership of several unknowns, based on your flow diagram.

Family Name	Functional Group	Family Name	Functional Group
Alkanes	R – H	Aldehydes	$\overset{\displaystyle O}{\underset{\displaystyle R-C-H}{\|}}$
Alkenes	$\underset{\displaystyle \|\ \ \|}{R-C=C-R}$	Ketones	$\overset{\displaystyle O}{\underset{\displaystyle R-C-R}{\|}}$
Alkynes	R – C ≡ C – R	Carboxylic Acids	$\overset{\displaystyle O}{\underset{\displaystyle R-C-OH}{\|}}$
Aromatics	(aromatic ring)	Esters	$\overset{\displaystyle O}{\underset{\displaystyle R-C-O-R}{\|}}$
Alcohols	R – OH	Amines	$\underset{\displaystyle \|}{R-\overset{\displaystyle ..}{N}-}$
Mercaptans	R – SH	Amides	$\overset{\displaystyle O}{\underset{\displaystyle R-C-N-}{\|}}$
Ethers	R – O – R		

Table 25–1: Organic family functional groups.

Investigation 1

Objective

To determine if smaller members of the alcohol, alkane, alkene, amine, and carboxylic acid families are soluble in water.

Safety Requirements

- Wear eye protection at all times when in the laboratory.

- Avoid contact with all chemicals. If skin contact occurs with any chemical, flood the area with water and then notify the instructor.

- Use a clean or provided dropper to dispense reagents.

- Use a clean and dry stirring rod.

- Do not contaminate the reagent supply bottles by returning unused or unwanted reagents.

- Dispose of all chemicals following the directions of your instructor.

- Keep all chemicals well away from flames, unless intentional.

Procedure

1. Select five 10×75 mm (4 mL) test tubes and place them in a holder.

2. Using masking tape or some other method, mark each test tube with the organic family test solution that will be placed in that tube.

3. Using the alcohol, alkane, alkene, amine, and carboxylic acid compounds provided, place approximately 10 drops of the representative organic family test solution in a test tube as shown in Figure 25-1.

4. Add 10 drops of DI water to each test tube containing the organic family test solutions.

5. Using a clean dry stirring rod each time, mix the contents of the test tube and note if it forms a clear solution (soluble), a cloudy solution (partially soluble), or separates into two layers (not soluble).

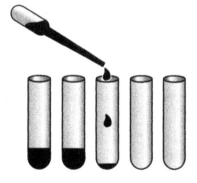

Figure 25-1: Use 10 drops of the organic test solution in a tube.

6. Make note of your observations.

7. Save the test tubes and contents of any organic families that appeared to be water soluble for Investigation 3.

Interpretation

1. Based on your observations, complete Flow Diagram 1 placing the organic family names in the appropriate boxes.

Flow Diagram 1

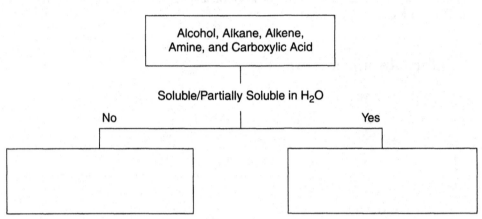

Investigation 2

Objective

To determine which of the water insoluble organic families decolorize a bromine solution.

Procedure

1. Using clean dry test tubes, place 10 drops of the water insoluble organic family test solution in separate test tubes.

2. Note the color of the Br_2 test solution. Add 3 drops of the test solution to each test tube. If the color remains, it is a negative test. If the color disappears, it is a positive test and indicates that it is an unsaturated organic compound.

3. Enter the names of the organic families that are not water soluble in the box at the top of the flow diagram and complete Flow Diagram 2, based on your findings.

Interpretation

Flow Diagram 2

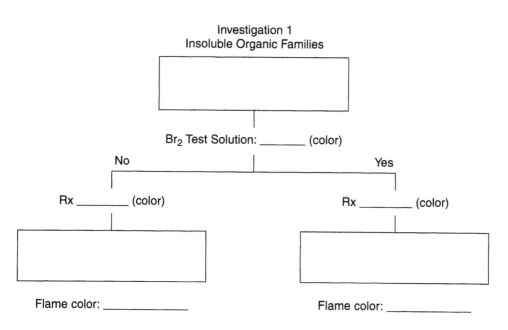

Investigation 1
Insoluble Organic Families

Br_2 Test Solution: _____ (color)

No   Yes

Rx _____ (color)   Rx _____ (color)

Flame color: _____   Flame color: _____

1. Which organic family(ies) is/are saturated?

2. Which organic family(ies) is/are unsaturated?

3. Clear your work area of all other chemicals. Using a clean dry evaporating dish, place 2 drops of one of the identified organic family test solutions. Using a match or burner, ignite the drops of liquid and note the color of its flame. Record the flame color below Flow Diagram 2.

4. After the evaporating dish has cooled, make sure it is again clean and dry and add 2 drops of the other organic family test solution identified. Using a match or burner, ignite the liquid and note its flame color. Record its flame color in Flow Diagram 2. (Remember: A blue flame confirms a saturated compound and a yellow flame confirms an unsaturated compound.)

Investigation 3

Objective

To develop a method to distinguish between water soluble organic families.

Procedure

1. Return to the test tubes of water soluble organic families identified and saved in Investigation I. Using a clean dry stirring rod, dip it into one of the test tubes containing one of the water soluble organic families and place the clinging drop of solution on a piece of red litmus paper. Repeat the procedure with a second drop on a piece of blue litmus paper. Note the color change, if any.

2. Repeat step 1 for each of the water soluble organic families.

3. Complete Flow Diagram 3 by placing the appropriate family name in each box, using the information gained.

Interpretation

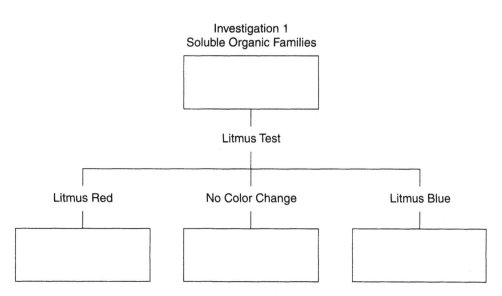

Flow Diagram 3

1. Which water soluble organic family results in a neutral solution? Write that family name in the box corresponding to "No Color Change."

2. Which water soluble organic family produces a solution that tests acidic (red) to litmus? Write that family name in the box corresponding to "Litmus Red."

3. Which water soluble organic family produces a solution that tests basic (blue) to litmus? Write the family name in the box corresponding to "Litmus Blue."

4. Confirm that the organic family that tested red to litmus can be neutralized by adding a base. Place two drops of this organic family test solution in a clean dry evaporating dish. Cautiously note its odor. Add two drops of 0.1 M NaOH solution. Be sure the drops mix with each other. Again, carefully note the odor. Describe any changes that result.

5. Confirm that the organic family that tested blue to litmus can be neutralized by adding an acid. Place two drops of this organic family test solution in a clean dry evaporating dish. Cautiously note its odor. Add two drops of 0.1 M HCl solution. Be sure the drops mix with each other. Again, carefully note the odor. Describe any change that results.

6. Confirm the organic family that gave no reaction to litmus by placing two drops of its test solution in a clean dry evaporating dish and adding two drops of chromic acid test solution. (*Caution*: This solution is very corrosive. If you should get any of it on your skin, wash immediately with water and have someone contact your instructor.) A change in color from orange to blue-green will confirm the presence of an alcohol. Record your observations.

Investigation 4

Objective

To identify Unknowns A, B, and C by organic family, using the Flow Diagrams developed.

Procedure

1. Following the directions of your instructor, empty, clean, and dry your test tubes.

2. Using similar methods to those used in the previous investigations, determine the organic family represented by Unknowns A, B, and C. (Note: An organic family may be used as an unknown more than once.)

3. Record the tests performed for each unknown solution and the results obtained.

Unknown A

Unknown B

Unknown C

Interpretation

1. In the spaces provided, indicate which organic family is represented by each of the unknowns.

   Unknown A _____

   Unknown B _____

   Unknown C _____

2. Indicate any confirmatory tests run and their results.

Report Sheet

Name: _____

Experiment 25

Date: _____ **Section:** _____

1. Using the information developed in Investigations 1, 2, and 3, complete the following Flow Diagram by writing the organic family names in the appropriate boxes.

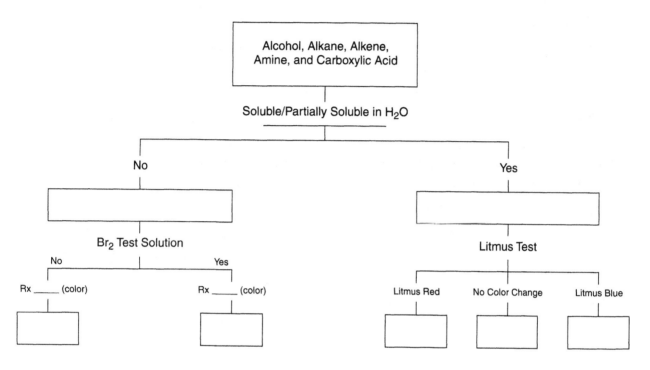

2. Complete Table 25-2, indicating the observed change for each of the confirmatory tests conducted.

Organic Family	Test Conducted	Observation
Alcohol		
Alkane		
Alkene		
Amine		
Carboxylic acid		

3. Identify Unknowns A, B, and C by organic family:

Unknown A _____

Unknown B _____

Unknown C _____

Extra Credit

As previously noted, as the length of the carbon chain increases, water solubility decreases. Consider the following two scenarios for organic family members with more than four carbons. Based on the test results, determine which organic family they represent.

1. Scenario I

An oily compound with a fishy odor does not dissolve in water, but when 0.1 M HCl is added the compound dissolves and the odor disappears. Later, an equal amount of 0.1 M NaOH is added and the fishy odor returns. To which organic family does this compound belong? Using a general formula from Table 25-1, write equations for adding both the HCl and NaOH, and explain why the odor returns.

2. Scenario II

A white crystalline powder does not dissolve in cold water, but does dissolve in warm water. When a drop of the warm water is tested, it turns blue litmus paper red. It was also discovered that when 0.1 M NaOH solution is added to the cold water, the white powder dissolves. Using a general formula from Table 25-1, write an equation for the addition of NaOH and explain why the crystals dissolve.

Experiment

26

How can you calculate the age of the Egyptian pyramids?

Simulating Nuclear Processes

Background Concepts

Radioactive isotopes are composed of atoms with unstable nuclei that will eventually disintegrate (decay). As the result of the disintegration, different kinds of detectable radiation are emitted. When a sample contains a large number of identical radioactive atoms, they do not all decay immediately or even at the same time. At any given moment, we are unable to predict which atom will decay next, or even how long it will be before it does. Nevertheless, when considering the total sample, there is a pattern to their decay. The probability that each atom has of decaying in the next second is the same. This results in exponential decay, in which the number of radioactive atoms decreases quickly at first and then slows in proportion to the number of radioactive atoms remaining.

Although we can't say which atom is going to decay at any instant, we can still discuss their overall decay rate. The *activity* of a radioactive substance is a measure of how many decays happen per second. An activity of 10 becquerel (10 Bq) means that there are 10 decays per second. A more common way to measure activity, however, is in curies, Ci, where 1 Ci = 3.70×10^{10} Bq.

Since each identical radioactive atom has the same probability of decaying each second, the number of decays that actually happen per second is a fraction, or a percentage, of the number of atoms remaining. This fraction can be calculated by using the *decay constant*, λ, in the following equation:

$$\frac{\text{number of decays}}{\text{second}} = \lambda \times (\text{number of radioactive atoms left})$$

For example, let's say that each atom has a 10% chance of decaying each second ($\lambda = 0.10/s$). Statistically, we would expect that of 1,000 atoms present at a given instant, 100 of them would decay in the next second.

$$\frac{\text{number of decays}}{\text{second}} = (0.10/s \times 1{,}000 \text{ atoms}) = 100 \text{ atoms/s}$$

The number of atoms remaining is constantly decreasing as more and more of them decay. The number of radioactive atoms remaining at any time, t, is given by the following function:

$$N = N_0 \, e^{-\lambda t}$$

$$N = N_0 \exp(-\lambda t)$$

where, N_0 is how many radioactive atoms you started with at $t = 0$.

The fewer radioactive atoms there are, the fewer atoms decay each second, and the more the decay process slows. It could take virtually forever for every radioactive atom to decay, so it doesn't make sense to talk about how long it takes for all the atoms to decay. Instead, we can talk about how long it takes for 50% of the radioactive atoms to decay, called the *half-life*, which is the time it takes for any given number of atoms to be reduced by 50%. The half-life, $T_{\frac{1}{2}}$, therefore, can be calculated by using the following formula:

$$T_{\frac{1}{2}} = \frac{0.0693}{\lambda}$$

Although dice do not undergo fission reactions, the probability that any particular number will be on its top after each roll is used in this experiment to simulate such a random nuclear event.

■ Investigation 1

Objectives

To monitor a statistical decay using dice, plot the decay graphically, compare the decay to a true exponential, and calculate the half-life of the decay.

Procedure

1. Place 100 dice on your tray. These represent 100 radioactive atoms.

2. Roll the dice and record the information, as detailed in steps a and b below, in Table 26-1. Each turn will represent one minute.

   a. Dice displaying "1's" on the top represent atoms that have decayed during that minute. Count and record the number of "1's" and then remove them from the tray

   b. Now count and record the number of atoms remaining on the tray.

3. Repeat step 2 until all the dice have decayed. Record the results of each roll in the table. (There may be some rolls [minutes] when no dice decay – record these minutes also.)

Minute	Atoms decayed this minute	Atoms remaining	Sec. B Calc.	Minute	Atoms decayed this minute	Atoms remaining	Sec. B Calc.
0	0	100		13			
1				14			
2				15			
3				16			
4				17			
5				18			
6				19			
7				20			
8				21			
9				22			
10				23			
11				24			
12				25			

Table 26–1: Dice for dice roll.

4. Open a new document in Excel®. In cell A1, label the first column "Time (mins)." Label the second column "Exponential." Label the third column "Atoms Remaining" as shown in Figure 26-1.

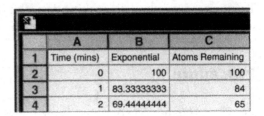

	A	B	C
1	Time (mins)	Exponential	Atoms Remaining
2	0	100	100
3	1	83.33333333	84
4	2	69.44444444	65

Figure 26-1

5. Fill in the first row as shown using "0 minutes," "100 Exponential," and "100 Atoms Remaining."

6. In cell A3, enter the formula =A2+1. In cell B3, enter the formula =B2*5/6. Select these two cells by dragging the cursor over them. COPY the contents of these two cells (you can do this by right clicking on them); select roughly the next 25 cells down of columns A and B with your cursor, and PASTE the formulas into them.

7. Enter your data from step 2 into Column C.

8. Select the cells from A1 down to C25. Using the "Insert Chart" option, create an XY scatter plot of the columns. The chart should have both "Exponential" and "Atoms Remaining" on the y-axis vs. "Time" on the x-axis. Refer to Lab 6 for the procedure to plot a graph using Excel®. Adjust your graph so that the "Exponential" (Column B) appears as a smooth line only, and your data (Column C) appear as point symbols. Print the graph.

9. In the 4th column of Table 26-1, perform the following calculation. Start with 100 atoms in the top row (minute 0 – write "100" into the table). Now, multiply that by 5/6. Write your answer in the cell below (minute 1). Then, multiply that answer by 5/6, and write it into the cell below that. Continue this way until you have filled up the whole column.

10. Plot the calculations from step 4 on Graph 26-1. Make your points small, then connect them with a smooth line, so that only the line is visible.

11. On top of the line, plot your data from the Atoms remaining columns in Table 26-1 as points that are large enough to be seen.

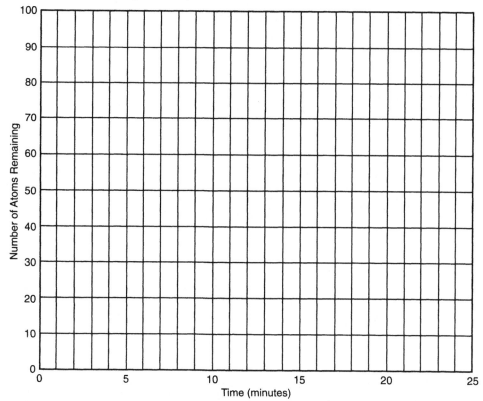

Graph 26–1: Decay of simulated atoms.

Interpretation

1. How well do your data points match the theoretical exponential curve?

2. Based on your graph, estimate how much time it takes to go from 100 to 50 atoms:

$$T_{\frac{1}{2}} = \underline{\hspace{2cm}}$$

3. Calculate the half-life of this decay using a different (convenient) starting point:

   a. Starting point:

   number of atoms is $N_0 =$ _____
   (pick a point with fewer than 100 atoms)

   time = _____

   b. Decay by 50%:

   number of atoms left is $\frac{1}{2} N_0 =$ _____

   time = _____

   c. Time elapsed between these two points

   time = _____

   d. Does this match the half-life from question 2?

   _____

4. Based on your graph, how long does it take the collection of atoms to go from 100% of its starting amount down to 25%?

5. How many times $T_{\frac{1}{2}}$ is this?

6. We can calculate the decay constant. Starting the decay at 100 dice means that $N_0 = 100$. In the lines below, fill in the value of time that it takes to reach $N = 50$ dice. (Remember to convert the time into seconds).

$$N = N_0\, e^{-\lambda t}$$

$$(50) = (100)e^{-\lambda(T_{\frac{1}{2}})}$$

$$\frac{50}{100} = e^{-\lambda(\underline{\quad})}$$

$$\frac{1}{2} = e^{-\lambda(\underline{\quad})}$$

$$\ln\left(\frac{1}{2}\right) = -\lambda(\underline{\quad\quad})$$

Based on this expression, calculate:

$$\lambda = \underline{\quad\quad}\ s^{-1}$$

7. How many decays are there per second, during the first minute?

$$\underline{\quad\quad}\ Bq$$

8. How many decays are there per second after one half-life?

$$\underline{\quad\quad}\ Bq$$

Is this half of your answer to question 7?

9. Express the answers from questions 7 and 8 in Curies.

10. For the theoretical curve, explain why each minute's count should be equal to 5/6 the previous value.

11. Not all dice are six-sided (see Figure 26-2). How would this decay be different if 12-sided dice were used?

Figure 26-2: Dice types.

12. How would this decay be different if both ones and twos were considered to be decayed atoms? Specify how the half-life would be affected.

Investigation 2

Objectives

To monitor the simulated decay of a first group of "mother" dice and monitor the simulated decay of a second set of "daughter" dice produced by the first decay.

Procedure

1. Count out 100 dice and place them in a bin. These will represent the "mother" atoms. Have a second, empty bin, for the "daughter" atoms.

2. Roll the dice in the first bin and the dice in the second bin. (The first minute, there won't be any dice in the second bin.) Let each turn represent one minute.

3. Remove all the dice that show a "1" from the daughter bin. These represent the atoms that have decayed into the final product. You may set them aside. (There won't be any for the first turn.)

4. Remove all the dice that show either a "1" or a "2" from the mother bin. Place these dice into the daughter bin. These represent the mother atoms that have decayed and transmuted into daughter atoms.

5. Finally, count the number of dice remaining in each bin, as well as the number of dice representing the final product, which have been taken out. Plot a point for each of these three categories on Graph 26-2. The 3 points indicate how many mother atoms and daughter atoms there are after the first minute, and how many atoms have been removed. You may use different symbols (O, Δ, ■) to identify the various points.

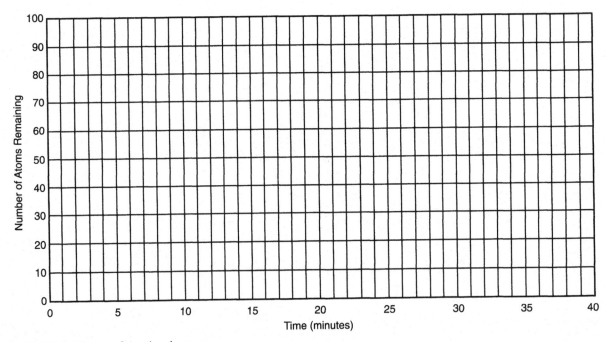

Graph 26–2: Decay of simulated atoms.

6. Repeat steps 2 – 5 for each successive minute until all the atoms have been removed from both bins.

7. For each of the three sets of data points, draw a smooth line through the points, and label each line to indicate whether it represents the mother atoms, daughter atoms, or final product.

Interpretation

1. How does the decay rate of the mother atoms compare to the decay rate of the daughter atoms?

2. Why would it be difficult to calculate the half-life of the daughter atoms based on the graph?

3. At roughly what time is the amount of product increasing the fastest?

4. Does this correspond to the time when the greatest number of daughter atoms are present?

5. For the mother atoms, what is the activity at the start, i.e., how many decays are there per second?

_____ Bq

6. For the mother atoms, what is the activity after one half-life?

_____ Bq

7. How are the values from questions 5 and 6 related?

Investigation 3

Objectives

To simulate an exponential growth using dice, plot the exponential growth on a graph, and calculate the doubling time of an exponential growth.

Procedure

1. You will be provided with roughly 100 dice. Place one of them in a bin.

2. Let each turn represent one minute.

3. Roll the dice in the bin. (At first, there will be only one). Count how many "1's" are showing. Add a die for every "1" showing.

4. Record how many dice are in the bin after each minute. Plot this number as a data point on Graph 26-3.

5. Repeat steps 3 and 4 until you have used up the 100 dice and no longer can add enough dice to the bin.

6. Draw a smooth line that passes reasonably close to each data point and illustrates the growth trend.

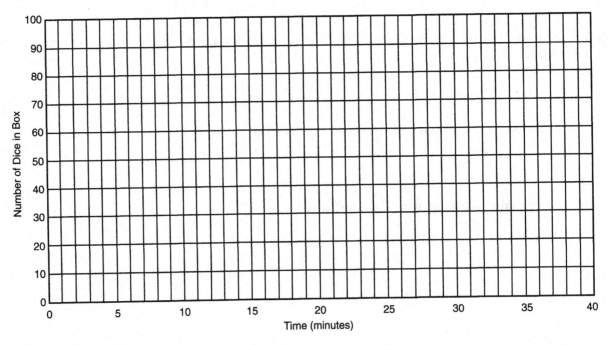

Graph 26–3: Simulation of exponential growth.

Interpretation

1. Based on your graph, estimate the following doubling times:

    a. How long does it take to go from 10 dice to 20?

    _____ minutes

    b. How long does it take to double from 20 to 40 dice?

    _____ minutes

    c. How long does it take to double from 40 to 80 dice?

    _____ minutes

    d. Are all of the times reasonably similar?

2. By what fraction or percentage does the number of dice grow each minute?

3. Based on the doubling times, let's estimate a "growth constant" λ, similar to the decay constant we did in Investigation 1 (Only here, there's no "−" in the exponential function: $N = N_0 \, e^{\lambda \, t}$. If we start at 10 atoms, then $N_0 = 10$. Fill in the time it takes to go from 10 to 20 atoms from question 1a.
   (Convert the time to seconds)

   doubling time = _____ seconds

   $$N = N_0 \, e^{\lambda \, t}$$

   $$(20) = (10)e^{\lambda(t_{\text{doubling}})}$$

   $$\frac{20}{10} = e^{\lambda(\underline{\quad})}$$

   $$2 = e^{\lambda(\underline{\quad})}$$

   $$\ln(2) = \lambda(\underline{\quad\quad})$$

   Based on this expression, calculate:

   $$\lambda = \underline{\quad\quad} \, s^{-1}$$

4. Based on the above exponential function and starting with 1 atom, how many atoms would you have after 1 hour?

5. In a nuclear reactor, when an atom undergoes fission, it emits neutrons that can trigger the decay of some of the surrounding atoms. If each atom causes only 1/6 of its neighbors to fission, how can this be dangerous?

Investigation 4 – Extra Credit

Objectives

To measure the half-life of an exponentially decaying voltage, calculate an appropriate decay constant for the voltage, and apply these numbers to calculate specific decay times.

Procedure

1. Your instructor will provide you with the following:
   - 1 Farad capacitor
   - galvanometer
   - 3 batteries
   - 10 K resistor
   - 33 ohm resistor
     (an analog voltmeter may be used as an alternative to the galvanometer and 10 K resistor).

2. Build the circuit as shown in Figure 26-3. Leave one of the battery wires unconnected as shown.

3. Touch the loose wire to the batteries, so that they charge the capacitor to a voltage of over 4 volts.

4. As you release the wire, the charges built up inside the capacitor will run out through the 33 ohm resistor. The voltage remaining across the capacitor will decay gradually. The galvanometer and 10 K resistor together form a voltmeter that can be used to monitor this decay.

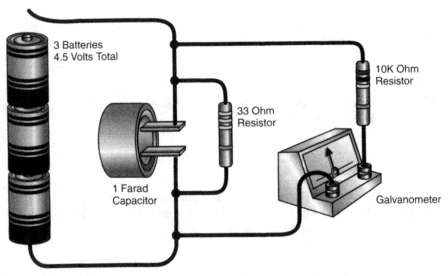

Figure 26-3

5. Monitor the capacitor's discharge. Using a stopwatch, measure how long it takes for the following voltage changes to happen. Record your results in Table 26-2.

Voltage drop	Time
From 4 V to 2 V	
From 2 V to 1 V	
From 3 V to 1.5 V	

Table 26–2: Capacitor voltage drop.

Interpretation

1. On average, what is the half-life of the capacitor's discharge?

2. How long would it take for the capacitor to discharge from 4 volts down to 0.5 volts? (How many half-lives is this?)

3. The decay of the voltage is exponential. The decay constant is represented by λ. If we start the decay at 4 volts, then $V_0 = 4$ V. In the lines below, fill in the time it takes to go from 4 volts to 2 volts:

$$V = V_0\, e^{-\lambda\, t}$$

$$(2\text{ V}) = (4\text{ V})e^{-\lambda(T_{\frac{1}{2}})}$$

$$\frac{2\text{ V}}{4\text{ V}} = e^{-\lambda(\underline{\quad})}$$

$$\frac{1}{2} = e^{-\lambda(\underline{\quad})}$$

$$\ln\left(\frac{1}{2}\right) = -\lambda(\underline{\quad})$$

Based on this expression, calculate:

$$\lambda = -\underline{\qquad}\ \text{s}^{-1}$$

4. Using the exponential function $V = V_0\, e^{-\lambda\, t}$, calculate how low the voltage would be after 30 seconds, starting at 4 volts. Use the value of λ you calculated above.

5. Using the same expression, calculate how long it would take the capacitor to discharge from 4 volts down to 0.001 volts.

$$V = V_0\, e^{-\lambda t}$$

$$(0.001\ \text{V}) = (4\ \text{V})e^{-\lambda t}$$

$$\frac{0.001\ \text{V}}{4\ \text{V}} = e^{-\lambda t}$$

$$\frac{0.001}{4} = e^{-\lambda t}$$

$$\ln\left(\frac{0.001}{4}\right) = -\lambda t$$

$$t = \underline{\hspace{2cm}}$$

Report Sheet

Experiment 26

1. Investigation 1

    a. On average, what was the half-life of the decay? _____

    b. What was the decay constant? _____

    c. Attach the graph from Investigation 1.

2. Investigation 2

    a. Attach the graph from Investigation 2.

3. Investigation 3

    a. What was the doubling time of the exponential growth? _____

    b. What was the "growth constant? _____

    c. Attach the graph from Investigation 3.

4. Investigation 4 (Extra Credit)

    a. On average, what was the half-life of the capacitor's discharge? _____

    b. What was its decay constant? _____

5. If the half-life is the time it takes for 50% of a radioactive substance to decay, why is it wrong to say that 100% of it will decay in a period of two half-lives?

6. What percentage of a radioactive substance will have decayed after two half-lives?

7. Living organisms are constantly exchanging carbon atoms through carbon dioxide in the air. Therefore, a small but steady amount of carbon-14, which is naturally radioactive, appears in the body of every organism. Because of this, for every gram of carbon atoms in the organism, there should be 16 decays happening per minute. The half-life of this decay is 5,730 years.

A mummified Egyptian cat is the subject of a study. It is found that a section of it containing a total of 10 grams of carbon produces only 8 decays per minute.

a. Express the present activity of one gram of the cat's carbon in becquerels.

_____ Bq

b. Of the carbon-14 atoms it was buried with, what fraction of them are still radioactive?

_____

c. If we assume that the cat died and stopped replenishing its supply of carbon-14 atoms when the pyramid was built, how old is the pyramid?

_____

Extra Credit

1. Using your data from Investigation 2, plot a graph of ln (number of atoms) vs. time. Show that when you plot the logarithm like this you get a straight line – until you start running out of dice.

Supplemental Exercises in Physics and Chemistry

1 – Significant Figures and Mathematical Operations

Possible Points = 40

I. Counting, Defined, and Measured Numbers

Counting Numbers are the cardinal numbers 1, 2, 3, 4, 5, 6, 7, 8, and 9 that are known exactly and may or may not have units. When you say "5 apples," you have exactly 5 apples.

Defined Numbers are numbers that look like measured numbers because they include units. In the statement "12 inches equals 1 foot," the numbers 12 and 1 are again exact because they are defined quantities.

Measured Numbers are the numbers obtained using a measuring instrument. Each measuring instrument limits the number of digits that can be read. If you measure the length of this page to within 0.1 cm (e.g. 27.7 cm), the true length is somewhere between 27.6 and 27.8 cm; that is, the length of this page is 27.7 ± 0.1 cm.

II. Significant Figures

Definition: Significant Figures are all numbers whose values are known exactly and, if they are the result of a measuring instrument, one estimated number is also included.

Rules for Determining Significant Figures

1. Counting and defined numbers have an unlimited number of significant figures.

   (5 apples and 12 inches = 1 foot both have an unlimited number of significant figures.)

2. All cardinal numbers that are read from a measuring instrument are significant.

   (27.7 cm and 185 lbs are both measurements with 3 significant figures.)

3. Zeros may or may not be significant.

   a. All zeros between cardinal numbers are significant.

      (205.06 g and 12,075 gal are both measurements containing 5 significant figures.)

   b. All zeros right of the decimal, but left of cardinal numbers are *not* significant.

      (0.00255 mm and 0.0203 miles are measurements that contain 3 significant figures.)

   c. All zeros right of cardinal numbers and right of the decimal are significant.

      (204.780 km and 5.04000 kg are both measurements containing 6 significant figures.)

   d. Zeros right of cardinal numbers, but left of the decimal, e.g., 25,000, may or may not be significant. The only way to prevent possible confusion is to put these numbers in scientific notational form (e.g., if it has 2 significant figures 2.5×10^4 or if it has 5 significant figures 2.5000×10^4).

For each of the following measurements, indicate the number of significant figures.

1. 34.56 g   _____

2. 0.5 ft   _____

3. 60.0057 in   _____

4. 95,000 lb   _____

5. 3,205 qts   _____

6. 203.060 meters   _____

7. 545,040 liters   _____

8. 2,555.008 fl oz   _____

9. 0.000540 inches   _____

10. 1,005,040 km   _____

III. Scientific Notation

1. Because science uses both very small and very large numbers, they are often expressed in scientific notational form. Use the following steps when you wish to express a number in notational form.

   a. Move the decimal point to the left or right enough places so the resulting number is between 1 and 10.

   b. Include all significant figures in the measurement.

   c. Indicate the number of places you have moved the decimal by the power on the 10.

   (54,500 feet = 5.45×10^4 ft or 0.0005670 mm = 5.670×10^{-4} mm)

One additional advantage of placing measurements in scientific notational form is that it eliminates the uncertainty of whether a zero is or is not significant.

Write each of the following measured values in correct scientific notational form.

1. 54.56 g   _____

2. 0.5 ft   _____

3. 60.0057 in   _____

4. 95,000 lb   _____

5. 8,205 qts   _____

6. 205.060 meters   _____

7. 545,040 liters   _____

8. 2,555.008 fl oz   _____

9. 0.000540 inches   _____

10. 1,005,040 km   _____

IV. Mathematical Operations and Significant Figures

Numbers with varying amounts of significant figures are frequently used in mathematical operations. There are two sets of rules involved in expressing your answers to the correct number of significant figures: one set for addition and subtraction, and the other for multiplication and division.

Rules For Rounding Numbers

1. When mathematical operations are performed, it is necessary to round the answer to the correct number of significant figures. The rules to use when rounding numbers are the following:

   a. When the first digit after those being retained is less than 5, simply drop the remaining numbers. (24.543 g rounded to 4 significant figures = 24.54 g)

b. When the first digit after those being retained is larger than 5, the last digit to be kept is increased by one. (456.468 cm rounded to 5 significant figures = 456.47 cm).

c. When the first digit after those being retained is a 5 and all others beyond it are zeros, then the last digit retained remains the same if it is an even number, or is increased by one number, if it is an odd number. (34.4500 cm would give 34.4 cm, but 34.3500 cm rounded to 3 significant figures = 34.4 cm)

d. When the first digit after those being retained is a 5 and all others beyond it are more than zeros, then the last retained digit is increased by one. (356,522 miles rounded to 3 significant figures = 357,000 miles)

Round the following numbers to the number of significant figures indicated.

Number	Significant Figures	Answer		Number	Significant Figures	Answer
1. 254.5764	4	_____		6. 54.5650	4	_____
2. 0.00542	2	_____		7. 254,456	2	_____
3. 205.5455	6	_____		8. 569.951	4	_____
4. 20.090468	5	_____		9. 2,875.0052	6	_____
5. 46.040500	5	_____		10. 54.89898	1	_____

Addition and Subtraction

1. The answer in an addition or subtraction should contain no more digits to the right of the decimal point than are in the quantity that has the least number of digits to the right of the decimal point.

2. Complete the operation indicated and then round the numbers in the answer to the proper number of significant figures.

In the following examples, the addition or subtraction is completed first and then the answers rounded to 5 significant figures.

Examples:

24.372 cm

72.21   cm

6.1488 cm
_____

102.7308 cm and rounds to 102.73 cm

254.5689 g

– 22.34   g
_____

212.2289 g and rounds to 212.23 g

Multiplication and Division

1. When the mathematical operation that is being performed is multiplication and/or division, the following rules must be followed:

a. The answer can have no more significant figures than the number used that contains the least number of significant figures.

b. The calculation is completed using all numbers as given, then the answer is rounded to the correct number of significant figures.

Complete all the following calculations and express your answer to the correct number of significant figures. If it becomes necessary to round the answer, use the rules previously given to determine the last digit in your answer.

1. $234 \times 25 \times 78.3$ =   _____

2. 0.00456×455.9 =   _____

3. $34.5 + 359.876 + 65.9$ =   _____

4. $89.897 - 67.5$ =   _____

5. $\dfrac{23.45 \times 76.2}{14.3 \times 5.5}$ =   _____

6. $\dfrac{0.00786 \times 0.0022}{35.46 \times 0.034}$ =   _____

7. $\dfrac{(23.5)^2 \times 88}{12.5 \times 76.25}$ =   _____

8. $\dfrac{(23.4 + 34) \times 2.23}{35.46 \times 0.034}$ =   _____

9. $234.56 + 45.3689 + 3452.3 + 0.254$ =   _____

10. $8{,}978.0080 - 345.05$ =   _____

2 – Metric Units and Unit Analysis I

Possible Points = 25

I. You need to only determine the correct set of units, and indicate if those units express: area [A], mass [M], length [L], speed [S] or volume [V].

1. $26 \text{ cm} - 6 \text{ cm}$ = 20 _____ [   ]

2. $3.14 \, (2 \text{ cm})^2 \, 6 \text{ cm}$ = 75.4 _____ [   ]

3. $\dfrac{50 \text{ miles}}{\text{hr}} \times \dfrac{5280 \text{ ft}}{1 \text{ mi}} \times \dfrac{12 \text{ in}}{1 \text{ ft}} \times \dfrac{2.54 \text{ cm}}{1 \text{ in}} \times \dfrac{1 \text{ m}}{100 \text{ cm}} \times \dfrac{1 \text{ km}}{1000 \text{ m}}$ = 81 _____ [   ]

4. $60 \text{ lb} \times \dfrac{454 \text{ g}}{1 \text{ lb}} \times \dfrac{1 \text{ kg}}{1000 \text{ g}}$ = 27.2 _____ [   ]

5. $36 \text{ in} \times \dfrac{2.54 \text{ cm}}{1 \text{ in}} \times 10 \text{ cm}$ = 914 _____ [   ]

II. Fill in the missing conversion factors, written in fractional form, that would complete the following conversion problems.

1. $30 \text{ yd} \times \dfrac{3 \text{ ft}}{1 \text{ yd}} \times \dfrac{?}{1 \text{ in}} \times \dfrac{2.54 \text{ cm}}{100 \text{ cm}} \times \dfrac{1 \text{ m}}{100 \text{ cm}} = 27.4 \text{ m}$ _____

2. $34 \text{ gal} \times \dfrac{4 \text{ qt}}{1 \text{ gal}} \times \ \ ? \ \ = 128 \text{ liters}$ _____

3. $\dfrac{55 \text{ mi}}{\text{hr}} \times \dfrac{5280 \text{ ft}}{1 \text{ mi}} \times \dfrac{12 \text{ in}}{1 \text{ ft}} \times \ ? \ \times \dfrac{1 \text{ m}}{100 \text{ cm}} \times \dfrac{1 \text{ km}}{1000 \text{ m}} = 88.5 \, \dfrac{\text{km}}{\text{hr}}$ _____

4. $327 \text{ in}^3 \times \dfrac{2.54 \text{ cm}}{1 \text{ in}} \times \dfrac{2.54 \text{ cm}}{1 \text{ in}} \times \dfrac{2.54 \text{ cm}}{1 \text{ in}} \times \dfrac{1 \text{ ml}}{1000 \text{ ml}} \times \dfrac{1 \text{ liter}}{?} = 5.4 \text{ liter}$ _____

5. $\dfrac{32 \text{ lb}}{\text{in}^2} \times \dfrac{454 \text{ g}}{1 \text{ lb}} \times \dfrac{1 \text{ kg}}{1000 \text{ g}} \times \dfrac{1 \text{ in}}{2.54 \text{ cm}} \times \ ? \ = 2.25 \, \dfrac{\text{kg}}{\text{cm}^2}$ _____

III. Make the following English/Metric conversions using the unit analysis method. (You may do your work below or on a separate sheet.)

1. 100 ft   =   _____ meters

2. 20 gal   =   _____ liters

3. 80 lbs   =   _____ grams

4. 6 in^2 =   _____ cm^2

5. 30 cm   =   _____ in

6. $\dfrac{30 \text{ gal}}{\text{hr}}$ = $\dfrac{\text{_____ liters}}{\text{min}}$

7. $\dfrac{30 \text{ lb}}{\text{in}^2}$ = $\dfrac{\text{_____ kg}}{\text{cm}^2}$

8. 10 miles   =   _____ km

9. $\dfrac{30 \text{ miles}}{\text{hr}}$ = $\dfrac{\text{_____ km}}{\text{hr}}$

10. 0.5 gal   =   _____ cc

3 – Metric Units and Unit Analysis II

Possible Points = 25

I. Correctly identify only the units that result from the following operations.

1. $36 \text{ in} \times \dfrac{1 \text{ ft}}{12 \text{ in}} \times \dfrac{1 \text{ yd}}{3 \text{ ft}}$ 　　　　　= 　　*** _____

2. $\dfrac{7.6 \text{ kg}}{\text{cm}^3} \times \dfrac{1 \text{ cm}^3}{1 \text{ ml}} \times \dfrac{1000 \text{ ml}}{1 \text{ liter}}$ 　　　　= 　　*** _____

3. $\dfrac{14 \text{ ft}}{\text{mo}} \times \dfrac{12 \text{ in}}{1 \text{ ft}} \times \dfrac{2.54 \text{ cm}}{1 \text{ in}} \times \dfrac{1 \text{ m}}{100 \text{ cm}} \times \dfrac{12 \text{ mo}}{1 \text{ yr}}$ 　= 　　*** _____

4. $12 \text{ ft} \times 14 \text{ ft} \times \dfrac{12 \text{ in}}{1 \text{ ft}} \times \dfrac{12 \text{ in}}{1 \text{ ft}} \times \dfrac{2.54 \text{ cm}}{1 \text{ in}} \times \dfrac{2.54 \text{ cm}}{1 \text{ in}}$ 　= 　　*** _____

5. $\dfrac{16 \text{ kg}}{\text{cm}^2} \times \dfrac{1000 \text{ g}}{1 \text{ kg}} \times \dfrac{1 \text{ lb}}{454 \text{ kg}} \times \dfrac{2.54 \text{ cm}}{1 \text{ in}} \times \dfrac{2.54 \text{ cm}}{1 \text{ in}}$ 　= 　　*** _____

II. Solve the following Metric/English conversions.

1. How many meters are there in 10 feet? 　　　　　　　　　 _____

2. How many grams are there in 150 lbs? 　　　　　　　　　 _____

3. How many quarts are there in 10 liters? 　　　　　　　　 _____

4. How many fluid ounces are there in a pint? 　　　　　　　 _____

5. How many cc are there in a gallon? 　　　　　　　　　　 _____

6. How many mm are equal to 4 inches? 　　　　　　　　　 _____

7. How many cents/kg would be equal to 50 cents/lb? 　　　　 _____

8. How many cents/gal would be equal to 130 cents/liter? 　　 _____

9. How many square meters are equal to 60 yards^2? 　　　　 _____

10. How many cm/sec are equal to 80 ft/min? 　　　　　　　 _____

III. Make the following Metric/Metric conversions.

1. 30 m   =   _____ cm

2. 5 liters   =   _____ mL

3. 10 mg   =   _____ g

4. 230 mL   =   _____ liters

5. 12 cm^2 =   _____ mm^2

6. 5 km   =   _____ m

7. 500 ml   =   _____ cm^3

8. 14 cm   =   _____ mm

9. 120 kg/m^2 =   _____ g/cm^2

10. 250 mg   =   _____ grams

4 – Density and Specific Gravity

Possible Points = 20

I. Solve each of the following problems in an organized way in the space provided.

1. A rock with a mass of 150.0 g was placed in a graduated cylinder that contained 20.0 mL of water. After the rock was added, the water level read 32.5 mL.

   a. What is the volume of the rock in mL? _____

   b. What is the volume of the rock in cc? _____

   c. What is the density of the rock in g/cc? _____

2. If a substance has the density of 2.3 g/cc:

   a. How many cc of this substance would be required to equal 100 g? _____

   b. Would this substance float in water? _____

3. Sulfuric acid, H_2SO_4, has a density of 1.86 g/mL. What is the mass (in grams) of 1 liter of sulfuric acid?

   _____

4. If 50 mL of a liquid weighs 220.0 g, what is the density (in g/mL) of the liquid?

   _____

5. Using unit analysis, convert the density of water (1.00 g/mL) into lb/gal.

   _____

6. If gold has the density of 19.3 g/cc, what would be its specific gravity? (Show method of calculation.)

   _____

7. If zinc metal, Zn, has a specific gravity = 7.14, What is the mass of a piece of zinc that has a volume of 350 cc?

   _____

5 – Temperature Conversion

Possible Points = 20

I. Change the following Celsius temperatures to the corresponding Fahrenheit temperatures.

1. 37°C   _____ °F

2. –40°C   _____ °F

3. 120°C   _____ °F

4. 0°C   _____ °F

5. –5°C   _____ °F

II. Change the following Fahrenheit temperatures to the corresponding Celsius temperatures.

1. 20°F   _____ °C

2. 100°F   _____ °C

3. 98.6°F   _____ °C

4. –10°F   _____ °C

5. 212°F   _____ °C

III. Change the following Celsius temperatures to Absolute (Kelvin) temperatures.

1. 50°C   _____ K

2. –35°C   _____ K

3. 280°C   _____ K

4. –250°C   _____ K

5. 37°C   _____ K

IV. Change the following temperatures to Absolute (Kelvin) temperatures.

1. 40°C   _____ K

2. –65°F   _____ K

3. 100°F   _____ K

4. –25°C   _____ K

5. 0°F   _____ K

6 – Atomic Structure and Periodic Table

Possible Points = *33*

I. Use the average whole number atomic mass listed on the periodic table, unless otherwise indicated, for each element when completing the following table.

Symbol	Atomic Number	Atomic Mass	Number of protons, p^+	Number of neutrons, n^o	Number of electrons, e^- K	L	M	N
Na								
	12							
	12	26						
					2	8	8	1
			6					
			8	10				
P								
Na^+								
		35	17		2	8	8	0
H^+	1	2	1	1				

NOTE: If there are no electrons in an energy level, place a <u>zero</u> in that space.

 Name: ______________________________

Date: ____________________ Section: __________

7 – Naming Binary Compounds

Possible Points = 30

Directions: This exercise is intended for use in conjunction with Appendix B in the *Physical Science: What the Technology Professional Needs to Know* textbook.

I. Write the correct names for each of the formulas indicated in items 1 – 15.

1. $HBr_{(aq)}$ ____________
2. $CaCl_2$ ____________
3. Al_2S_3 ____________
4. KI ____________
5. HCl ____________
6. NH_4F ____________
7. $Ba(OH)_2$ ____________
8. $NaCN$ ____________
9. $FeBr_3$ ____________
10. CS_2 ____________
11. SiO_2 ____________
12. Cu_2O ____________
13. N_2O_4 ____________
14. FeO ____________
15. $(NH_4)_2S$ ____________

II. Write the correct formulas for the compounds named in items 1 – 15.

 1. Sodium sulfide   _____

 2. Iron(III) oxide   _____

 3. Aluminum chloride   _____

 4. Ammonium hydroxide   _____

 5. Calcium bromide   _____

 6. Diphosphorous pentoxide   _____

 7. Carbon tetrachloride   _____

 8. Cupric sulfide   _____

 9. Hydroiodic acid   _____

10. Lead(II) oxide   _____

11. Barium peroxide   _____

12. Radium chloride   _____

13. Antimony(V) fluoride   _____

14. Hydrosulfuric acid   _____

15. Lithium cyanide   _____

Name: ______________________

Date: ______________ **Section:** ________

8 – Naming Ternary Compounds

Possible Points = 30

Directions: This exercise is intended for use in conjunction with Appendix B in the *Physical Science: What the Technology Professional Needs to Know* textbook.

I. Write the correct names for each of the formulas indicated in items 1 – 15.

1. $NaNO_3$ ____________
2. $(NH_4)_2SO_4$ ____________
3. $Al_2(CO_3)_3$ ____________
4. H_2SO_4 ____________
5. $FePO_4$ ____________
6. $KMnO_4$ ____________
7. $HClO$ ____________
8. $Ba(HCO_3)_2$ ____________
9. $Mg(NO_3)_2$ ____________
10. H_2CO_3 ____________
11. $Cu_3(PO_4)_2$ ____________
12. K_2SO_3 ____________
13. $AgNO_3$ ____________
14. $NaClO_3$ ____________
15. $FeSO_4$ ____________

II. Write the correct formulas for the compounds named in items 1 – 15.

1. Sodium sulfate   _____

2. Iron(III) carbonate   _____

3. Aluminum chlorate   _____

4. Ammonium phosphate   _____

5. Calcium hypobromite   _____

6. Nitric acid   _____

7. Nitrous acid   _____

8. Cupric sulfate   _____

9. Ferric permanganate   _____

10. Lead(II) bicarbonate   _____

11. Barium oxalate   _____

12. Magnesium nitrate   _____

13. Antimony(V) perchlorate   _____

14. Sulfurous acid   _____

15. Sodium hydrogen phosphate   _____

Name: _____

Date: _____ Section: _____

9 – Naming Inorganic Compounds

Possible Points = 30

Directions: This exercise is intended for use in conjunction with Appendix B in the *Physical Science: What the Technology Professional Needs to Know* textbook.

I. Write the correct names for each of the formulas indicated in items 1 – 15.

1. CO _____

2. $KHCO_3$ _____

3. SnO_2 _____

4. $CaSO_4$ _____

5. HNO_3 _____

6. $Zn(NO_2)_2$ _____

7. $HClO_3$ _____

8. $BiBr_3$ _____

9. HgI _____

10. NH_4NO_3 _____

11. $CuCr_2O_7$ _____

12. $Co(HSO_4)_2$ _____

13. SnF_2 _____

14. $AgClO_3$ _____

15. $HCl_{(aq)}$ _____

II. Write the correct formulas for the compounds named in items 1 – 15.

1. Bismuth(III) sulfate   _____

2. Zinc chromate   _____

3. Hydrosulfric acid   _____

4. Calcium hydride   _____

5. Lead(II) bicarbonate   _____

6. Ferric acetate   _____

7. Carbon disulfide   _____

8. Carbonic acid   _____

9. Manganese(IV) sulfide   _____

10. Copper(I) hydroxide   _____

11. Sodium perchlorate   _____

12. Sulfur trioxide   _____

13. Aluminum phosphate   _____

14. Potassium cyanide   _____

15. Nickel(II) iodide   _____

10 – Completing and Balancing Chemical Equations

Possible Points = 20

Statement: There are thousands upon thousands of different chemical reactions. To aid the study of chemical reactions, they can be subdivided into several different types. In this assignment, four of these types have been included. Complete and balance each reaction, using the type heading as a general model.

I. **Combination Type:** $A + B \longrightarrow AB$

1. $H_2 + O_2 \longrightarrow$ _____

2. $Ca + O_2 \longrightarrow$ _____

3. $K + O_2 \longrightarrow$ _____

4. $H_2 + Br_2 \longrightarrow$ _____

5. $Al + Cl_2 \longrightarrow$ _____

6. $Na + Cl_2 \longrightarrow$ _____

7. $NH_3 + H_2O \longrightarrow$ _____

8. $NH_3 + HCl \longrightarrow$ _____

9. $CaO + H_2O \longrightarrow$ _____

10. $SO_3 + H_2O \longrightarrow$ _____

II. **Decomposition Type:** $AB \longrightarrow A + B$

1. $HgO \longrightarrow$ _____ + _____

2. $NaClO_3 \longrightarrow$ _____ + _____

3. $H_2CO_3 \longrightarrow$ _____ + _____

4. $NH_4OH \longrightarrow$ _____ + _____

5. $H_2O \longrightarrow$ _____ + _____

6. $NaCl \longrightarrow$ _____ + _____

7. $CaCO_3 \longrightarrow$ _____ + _____

8. $H_2CO_3 \longrightarrow$ _____ + _____

9. $CuSO_4 \cdot 5H_2O \longrightarrow$ _____ + _____

10. $H_2O_2 \longrightarrow$ _____ + _____

III. Single Replacement Type: AB + C $\longrightarrow$ CB + A or AB + D $\longrightarrow$ AD + B

1. $CuO + H_2 \longrightarrow$ _____ + _____

2. $HCl + Al \longrightarrow$ _____ + _____

3. $HCl + Mg \longrightarrow$ _____ + _____

4. $H_2SO_4 + Zn \longrightarrow$ _____ + _____

5. $KBr + Cl_2 \longrightarrow$ _____ + _____

6. $Al + H_2SO_4 \longrightarrow$ _____ + _____

7. $NaI + Br_2 \longrightarrow$ _____ + _____

8. $Cu(NO_3)_2 + Zn \longrightarrow$ _____ + _____

9. $Pb(C_2H_3O_2)_2 + Zn \longrightarrow$ _____ + _____

10. $CuCl_2 + Mg \longrightarrow$ _____ + _____

IV. Double Replacement Type: AB + CD $\longrightarrow$ AD + CB

1. $HCl + NaOH \longrightarrow$ _____ + _____

2. $H_2SO_4 + KOH \longrightarrow$ _____ + _____

3. $BaCl_2 + Na_2SO_4 \longrightarrow$ _____ + _____

4. $Ba(OH)_2 + HCl \longrightarrow$ _____ + _____

5. $AlCl_3 + AgNO_3 \longrightarrow$ _____ + _____

6. $Fe_2O_3 + HCl \longrightarrow$ _____ + _____

7. $NaHCO_3 + HCl \longrightarrow$ _____ + _____

8. $Pb(NO_3)_2 + NaCl \longrightarrow$ _____ + _____

9. $H_3PO_4 + Ca(OH)_2 \longrightarrow$ _____ + _____

10. $HC_2H_3O_2 + NH_4OH \longrightarrow$ _____ + _____

Reaction Types

(To be used in conjunction with Exercise 10 Completing and Balancing Equations)

I. Combination Type: A + B ⟶ AB

1. metal + oxygen ⟶ metal oxide

2. nonmetal + oxygen ⟶ nonmetal oxide

3. metal + nonmetal ⟶ salt

4. a soluble metal oxide + water ⟶ base

5. nonmetal oxide + water ⟶ an oxyacid

6. metal oxide + nonmetal oxide ⟶ salt

7. anhydrous salt + water ⟶ hydrate

8. ammonia + binary acid ⟶ ammonium salt

II. Decomposition Type: AB ⟶ A + B

1. heavy metal oxide ⟶ metal + oxygen

2. metal chlorate ⟶ metal chloride + oxygen

3. metallic nitrates ⟶ metal nitrites + oxygen

4. electrolysis of compounds ⟶ element + element

5. hydrates ⟶ anhydrous salt + water

6. metallic carbonate ⟶ metallic oxide + CO_2

7. metallic bicarbonate ⟶ metallic carbonate + H_2O + CO_2

8. peroxides ⟶ oxides + oxygen

III. Single Replacement Type: AB + C ⟶ CB + A <u>or</u> AB + D ⟶ AD + B

1. nonmetal activity F_2> Cl_2> Br_2> I_2

   salt $_1$ + free halogen $_1$ ⟶ salt $_2$ + free halogen $_2$

2. metal activity series in textbook

   salt $_1$ + more reactive metal ⟶ salt $_2$ + free metal

3. very active metals + water ⟶ metal hydroxide + hydrogen gas

4. active metals + acid ⟶ salt + hydrogen gas

IV. **Double Replacement Type:** $AB + CD \longrightarrow AD + CB$

   1. acid + base $\longrightarrow$ salt + water

   2. metal carbonate + acid $\longrightarrow$ salt + water + CO_2

   3. metal bicarbonate + acid $\longrightarrow$ salt + water + CO_2

   4. metal oxide + acid $\longrightarrow$ salt + water

   5. salt + base $\longrightarrow$ insoluble base + salt

   6. salt $_1$ + salt $_2$ $\longrightarrow$ insoluble salt $_3$ + salt $_4$

   7. ammonium salt + base $\longrightarrow$ salt + ammonia + water

   8. metallic sulfite + acid $\longrightarrow$ salt + SO_2 + water

   9. salt + acid $\longrightarrow$ insoluble salt + acid

11 – Speed and Acceleration

Possible Points = 20

$$\text{speed} = \frac{\text{distance}}{\text{time}} \qquad \text{acceleration} = \frac{\text{change in velocity}}{\text{time}} \qquad \begin{aligned} \text{force} &= \text{mass} \times \text{acceleration} \\ \text{weight} &= \text{mass} \times \text{gravity} \\ g &= 9.8 \text{ m/s}^2 \end{aligned}$$

1. A net force of 5.0 N is applied to an object that has a mass of 1.75 kg. What acceleration is produced?

2. A wagon's speed uniformly increases from 3.5 m/s to 5.8 m/s in 32 seconds.

   a. What is the average speed of the wagon during those 32 seconds?

   b. What is the distance traveled by the wagon during those 32 seconds

   c. hat is the distance traveled by the wagon during those 32 seconds?

   d. What is the acceleration of the wagon?

3. An object at rest is accelerated at a rate of 50 cm/sec^2. Determine the amount of time required to reach a speed of 80 cm/sec.

4. A 2.5 kg rock is dropped into a mine shaft and strikes water in 1.5 seconds. Ignoring the time for the sound to travel, how deep is the shaft?

5. The engineer on a train traveling at 30 m/s applies the breaks and brings the train to a uniform stop in 44 seconds. What is the acceleration of the train?

12 – Momentum and Friction

Possible Points = 20

$$KE = \tfrac{1}{2}mv^2 \qquad p = m\,v \qquad m_1 v_1 + m_2 v_2 = m_1 v_1 + m_2 v_2$$

$$f = \mu_K\,(mg) \qquad g = 9.8 \text{ m/s}^2 \text{ or } 9.8 \text{ N/kg}$$

1. A bowling ball has a mass of 7.2 kg and a speed of 2.5 m/s.

    a. What is its kinetic energy?

    b. What is its momentum?

2. A force of 3,000 N is required to start a sled whose mass is 800 kg, while a force of 1,200 N is sufficient to keep it moving once it is started.

    a. What is the coefficient of static friction?

    b. What is the coefficient of kinetic friction?

3. A 1,455 kg car traveling at a speed of 19 m/s strikes a wall and comes to rest after traveling another 12 cm. What is the average force exerted on the wall by the car?

4. A 15.0 g inelastic mass is moving with a velocity of 30 cm/s north when it collides with a 5.0 gram inelastic mass traveling at 70 cm/s south. If they stick together upon impact, what will be their resultant direction and velocity?

5. A ball with a mass of 0.5 kg is thrown against a wall. When it strikes the wall it is moving horizontally to the right at 30 m/s, and it rebounds horizontally to the left at 25 m/s. What is the impulse of the force exerted on the ball by the wall?

Name: _____

Date: _____ Section: _____

13 – Simple Machines

Possible Points = 20

$$F_G = mg \qquad\qquad \text{work} = F \times d \qquad\qquad F_i \times d_i = F_o \times d_o$$

$$\text{A.M.A.} = \frac{\text{force exerted by machine}}{\text{force exerted by you}} \qquad \text{I.M.A.} = \frac{\text{distance you pull}}{\text{distance load goes up}} \qquad \text{efficiency} = \frac{\text{A.M.A.}}{\text{I.M.A.}} \times 100\%$$

1. How much work is done when you lift a 550 kg object 10.0 m?

2. A pulley system is used to lift a 450 N block a distance of 3.0 m by the application of a 70 N force over a distance of 25.0 m.

   a. What is the actual mechanical advantage (A.M.A.) of the system?

   b. What is the ideal mechanical advantage (I.M.A.) of the system?

   c. What is the efficiency of the system?

3. A boy and his sled have a mass of 25 kg. The snow exerts a frictional force of 50 N. How much work is required to pull the boy and his sled 100 m up a 32° slope?

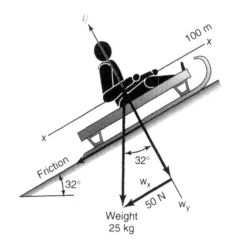

4. A chain hoist is used to lift a 300 kg block of stone a distance of 3.0 m by applying a force through a distance of 30 m. What is the mechanical advantage of the system?

5. An axle with a 5.0 cm radius is connected to a cogwheel with a 60 cm radius, as shown. What force, ignoring friction, must be applied to the cogwheel to wind the cable around the axle and lift a 200 kg load? (Hint: The axle and cogwheel circumferences can be found by using the formula $c = 2\pi r$).

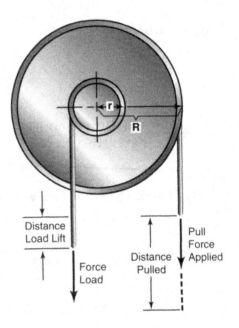

Distance
Load Lift

Force
Load

Distance
Pulled

Pull
Force
Applied

14 – Gas Laws

Possible Points = 24

Directions: Solve each of the following problems in an organized fashion.

1. The volume of a gas is 800 liters at 750 mm Hg (torr) and 20°C. What volume in liters will it occupy at 710 torr and 20°C?

2. The volume of a gas is 600 liters at 12°C. What volume will it occupy at 0°C, if the pressure remains constant?

3. A 550 mL sample of gas is collected at 400 torr and a temperature of 273°C. What volume would this gas occupy at STP (760 torr and 0°C)?

4. If 350 mL of a gas at 25°C was cooled to –40°C, pressure remaining constant, what will be the new volume?

5. If 1,500 mL of H_2 gas at 550 torr and 25.5°C were changed to STP, what would be its new volume?

6. If 625 mL of gas at 55°C is heated until its volume becomes 1,500 mL, what will be its temperature (°C) assuming that the pressure remains constant?

7. If 300 mL of gas at 0°C were to have its volume cut in half by decreasing the temperature, what temperature would be necessary? ($P = k$)

8. If 500 mL of gas at 500 torr were to have its volume cut in half by increasing the pressure, what pressure would be necessary? ($t = k$)

9. If 350 cc of an air/gasoline mixture were trapped at 30°C and 760 torr, what would be the pressure of the mixture if its volume were reduced to one-eighth of its original volume (compression ratio of 8:1) and the temperature increased to 60°C?

10. A 6.00 liter sample of Kr gas at 0.500 atm and 400 K is compressed until the pressure is 1.500 atm; the temperature is then changed at constant pressure until the volume is 9.00 liters. What is the new temperature (°C)?

 Name: ____________________

Date: ______________ Section: ________

15 – Calculating Molar Masses

Possible Points = 20

Atoms have definite masses, commonly called atomic mass. The formula of a compound indicates what atoms the compound contains and the number of each kind of atom. Hence, the formula mass can be determined by adding up all of the individual atomic masses present in a formula.

I. Find the formula mass of the following compounds; include units of either u/molecule or grams/mole on each answer.

1. HCl __________
2. KBr __________
3. $NaNO_3$ __________
4. $KMnO_4$ __________
5. $Ca(OH)_2$ __________
6. $NaHCO_3$ __________
7. $FeCl_3$ __________
8. CH_3CH_2OH __________
9. $Mg(NO_3)_2$ __________
10. NH_3 __________
11. H_2SO_4 __________
12. K_2SO_3 __________
13. $(NH_4)_3PO_4$ __________
14. $KHCO_3$ __________
15. $Li_2C_2O_4$ __________
16. $C_{12}H_{22}O_{11}$ __________
17. $Na_2Cr_2O_7$ __________
18. $KClO_3$ __________
19. $HC_2H_3O_2$ __________
20. Br_2 __________

16 – Numbers and Moles

Possible Points = 25

I. The mole can be compared to the dozen. Each word indicates that the items being counted are being grouped; the mole in groups of 6.02×10^{23} and the dozen in groups of 12. Stated in a slightly different way, 1 mole of anything is equal to 6.02×10^{23} of them. (1 mole = 6.02×10^{23})

II. The number of moles of a pure substance can also be converted into its mass by using that substance's formula mass. (Carbon 12 u/atom and therefore is 12 g/mole.)

Examples:

Carbon: 1 mole C = 6.02×10^{23} C atoms

                 1 mole C = 12 g C

Water: 1 mole H_2O = 6.02×10^{23} H_2O molecules

                 1 mole H_2O = 18 g H_2O

III. Calculate the correct number or mass for each of the following:

1. How many dollars would be in a mole of dollars?   _____

2. How many donuts would be in a mole of donuts?   _____

3. How many chickens would be in a mole of chickens?   _____

4. How many legs would be in a mole of chickens?   _____

5. How many atoms would be in a mole of atoms?   _____

6. How many molecules would be in a mole of molecules?   _____

7. How many molecules would be in a mole of O_2?   _____

8. How many atoms of oxygen would be in a mole of O_2?   _____

9. How many atoms of oxygen would be in a mole of H_2O?   _____

10. How many molecules would be in a mole of NH_3?   _____

11. How many atoms of hydrogen would be in a mole of NH_3?   _____

12. How many molecules would be in 0.5 moles of CH_4? _____

13. How many atoms of hydrogen would be in 2 moles of CH_4? _____

14. How many molecules would there be in 20 moles of H_2SO_4? _____

15. What is the mass of 1 mole of H_2? _____

16. What is the mass of 2 moles of H_2? _____

17. What is the mass of 0.5 moles of H_2? _____

18. What is the mass of 2 moles of NH_3? _____

19. What is the mass of 10 moles of HNO_3? _____

20. What is the mass of 6.02×10^{23} molecules of HNO_3? _____

21. How many molecules of HNO_3 are there in 21 g of HNO_3? _____

22. How many moles of HNO_3 would be in 126 g of HNO_3? _____

23. How many atoms of hydrogen would be in 63 g of HNO_3? _____

24. How many atoms of oxygen would be in 63 g of HNO_3? _____

25. How many atoms of all kinds would be in 63 g of HNO_3? _____

17 – Moles and Mass

Possible Points = 25

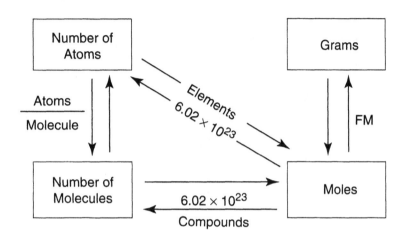

I. From the information given, calculate the number of moles of each substance represented.

1. 196 grams of H_2SO_4 _____

2. 1.96 grams of H_2SO_4 _____

3. 684 grams of $Al_2(SO_4)_3$ _____

4. 34.2 grams of $C_{12}H_{22}O_{11}$ _____

5. 7.8   grams of $Al(OH)_3$ _____

6. 6.02×10^{23} atoms of sulfur   _____

7. 1.20×10^{23} molecules of HCl   _____

8. 3.01×10^{23} atoms of Zn   _____

9. 1.50×10^{23} molecules of H_2CO_3 _____

10. 5.00×10^{23} molecules of $KHCO_3$ _____

II. Complete the following Table from the information given:

Formula for substance	Number of grams	Number of moles	Number of molecules	Number of atoms
$Cu(OH)_2$	100.0			
NaBr		2.0		
MgS				1.2×10^{24}
Br_2			6.02×10^{21}	
$KMnO_4$	63.2			

18 – Mole-Mole and Mass-Mass Stoichiometry

Possible Points = 25

I. Mole – Mole

	2 Al	+	6 HCl	$\longrightarrow$	2 AlCl$_3$	+	3 H$_2$
Mole Ratio:	$\dfrac{\text{2 moles}}{\text{____ moles}}$		$\dfrac{\text{6 moles}}{\text{____ moles}}$		$\dfrac{\text{2 moles}}{\text{____ moles}}$		$\dfrac{\text{3 moles}}{\text{____ moles}}$
	$\dfrac{\text{____ moles}}{\text{____ moles}}$		$\dfrac{\text{30 moles}}{\text{____ moles}}$		$\dfrac{\text{____ moles}}{\text{____ moles}}$		$\dfrac{\text{____ moles}}{\text{____ moles}}$
	$\dfrac{\text{____ moles}}{\text{____ moles}}$		$\dfrac{\text{____ moles}}{\text{____ moles}}$		$\dfrac{\text{____ moles}}{\text{____ moles}}$		$\dfrac{\text{2 moles}}{\text{____ moles}}$

II. Mass – Mass

	2 C$_4$H$_{10}$	+	13 O$_2$	$\longrightarrow$	8 CO$_2$	+	10 H$_2$O
Mole Ratio:	$\dfrac{\text{____ moles}}{\text{116 grams}}$		$\dfrac{\text{____ moles}}{\text{____ grams}}$		$\dfrac{\text{____ moles}}{\text{____ grams}}$		$\dfrac{\text{____ moles}}{\text{____ grams}}$
	$\dfrac{\text{10 moles}}{\text{____ grams}}$		$\dfrac{\text{____ moles}}{\text{____ grams}}$		$\dfrac{\text{____ moles}}{\text{____ grams}}$		$\dfrac{\text{____ moles}}{\text{____ grams}}$
	$\dfrac{\text{____ moles}}{\text{____ grams}}$		$\dfrac{\text{____ moles}}{\text{____ grams}}$		$\dfrac{\text{____ moles}}{\text{____ grams}}$		$\dfrac{\text{____ moles}}{\text{54 grams}}$

19 – Solution Concentration Percent and Dilution

Possible Points = 10

I. Definitions:

Concentration – the number of things per given unit of volume.

$$10\% \ (mass/vol) \ = \ 10 \ grams/100 \ mL$$

$$10 \ M \ = \ 10 \ moles/liter$$

1. How many grams of NaOH would be required to prepare 400 mL of 6% solution?

2. How many mL of a 3% glucose solution would be required to obtain 180 g of glucose?

3. If you wished to dilute 1 liter of a 3% glucose solution and make the resulting solution 2%, what would be the final volume of 2% solution?

4. A physiological saline solution is equivalent to 0.9% NaCl. (This solution has the same osmotic pressure as a red blood cell.) How many grams of NaCl would be required to make 500 mL of this solution?

5. If you wished to dilute a 2% NaCl solution to make 500 mL of 0.9% NaCl solution, how many mL of the 2% solution would you need?

20 – Molarity

Possible Points = 20

I. Definitions and Formulas:

Molarity – the number of moles of solute per liter of solution.

$$M = \frac{moles}{liter} \qquad M = \frac{grams}{formula\ mass \times liter}$$

II. Solve each of the following problems in an organized fashion. Use a separate sheet of paper if necessary.

1. Calculate the molarity of a solution prepared by dissolving 6 moles NaOH in sufficient water to make 1.5 liters of solution.

2. Calculate the volume of 6 M H_2SO_4 solution that would be required to give 3.00 g of H_2SO_4.

3. Calculate the volume of solution that would be required to give 16 moles of NaCl if the solution used is 12 M.

4. If you wished to dilute 300 mL of 6 M and make the resulting solution 4 M, how much water would you have to add?

5. What volume of solution would be required to make 3 moles of HCl into a solution that is 0.5 M?

6. How many grams of $Ca(OH)_2$ would be required to prepare 200 mL of a 6 M solution?

7. What is the formula mass of a substance if 49.0 g of it dissolved in 500 mL made a solution that is 1.0 M?

8. Find the molarity of a solution prepared by diluting 50 liters of a 7.5 M NaOH solution with an additional 200 liters of H_2O.

9. What would be the molarity of a solution made by adding 300 mL of 6 M HNO_3 to 200 mL of 4 M HNO_3?

10. Calculate the molarity of a 10%(mass/vol) KNO_3 solution.

21 – Logarithms and pH

Possible Points = 40

I. Introduction

Aqueous solutions are of particular importance because all living organisms use water as their primary solvent. Various substances dissolve in the water giving the resulting solutions their different characteristics. There are times when hydrogen ions in solution act as catalysts for reactions that occur in living materials. (Remember that catalysts can greatly change the rate of chemical reactions without being used up.) Very small changes in the concentration of hydrogen ions can have a very large effect on the speed of the reactions.

In 1907, Sörenson developed a mathematical formula to convert small concentrations of hydrogen ions into large numbers that would be easier for everyone to use and understand. These numbers are called the pH of the solution. The use of a pH number is, therefore, just another way to express the concentration of hydrogen ions available in the solution.

In this assignment, a method that can be used to convert the amount of hydrogen ions, H^+, given in moles/liter (M) to pH will be presented.

II. Logarithms

In mathematical terms, Sörenson's formula is $pH = -\log_{10}[H^+]$. The part of this formula that may be new to you is the $\log_{10}$ part, which stands for logarithm to the base 10. What this means is the power to which you must raise the number 10 to equal the number.

Examples: $100 = 1.00 \times 10^2$ therefore, $\log_{10} 100 = 2.0000$

$500 = 5.00 \times 10^2$ therefore, $\log_{10} 500 = 2.69897$

$0.0500 = 5.00 \times 10^{-2}$ therefore, $\log_{10} 0.0500 = -1.30103$

It should be noted that in the last example, 0.0500 is a number that is less than one. The logarithm for any number that is less than one, is a negative number.

III. Electronic Calculators

Electronic calculators have made the determination of logs (the common expression used for logarithms₁₀) an easy process. First, you need to have a calculator that is of the "scientific" variety. (This means that the calculator will have keys on it with notations like LOG, 10^x, EXP or EE, ln, etc.) Assuming your calculator is of this type, let's try the following example:

Examples: To determine the log of 500 (log 500 = _____), enter 500, then press the "LOG" key. Your display should read 2.69897, which may be rounded off to 2.6990.

To determine the log of 0.00500 (log 0.00500 = _____), enter 0.00500, then press the "LOG" key. The display should read –2.3010 (rounded).

Another method for entering this number is in the scientific notational form. The number 0.00500 can also be written as 5.00 ¥ 10–3. Enter in your calculator, 5.00 then push the "EE" or "EXP" key. This moves the 5.00 over and makes room on the right for the power of 10. Next, press the "3" key, followed by the "±" key. Now you are ready to press the "LOG" key, and the answer should again read –2.3010.

1. For each of the following, express the number in scientific notational form on the first blank, and then determine the log of the number, rounded to the fourth number after the decimal, and place it on the second blank.

   a. log 350.0   =   log _____ = _____

   b. log 0.0025   =   log _____ = _____

   c. log 0.0000000560   =   log _____ = _____

   d. log 0.00000000245   =   log _____ = _____

   e. log 0.00000000000957   =   log _____ = _____

IV. Equilibrium Constant Of Water, K_w

Now that we have a basic idea of how to convert a number into its scientific notational form and then into a log of that number, let's turn our attention to the equilibrium that exists in aqueous solutions.

The equation $H_2O \longrightarrow H^+ + OH^-$, represents the fact that at 25°C a small percentage of water molecules get broken into hydrogen ions and hydroxide ions. In the next second, these ions may again combine and become water molecules but in some other place in the solution another water molecule will be breaking apart. Chemists have been able to measure the concentration of the hydrogen and hydroxide ions and have determined their concentration, at this temperature, to be 1.00×10^{-7} moles/liter. Stated another way, $[H^+] = 1.00 \times 10^{-7}$ and $[OH^-] = 1.00 \times 10^{-7}$, where [ ] indicates that the concentration is being expressed in moles/liter (M).

Since the $[H^+]$ and $[OH^-]$ ion concentrations are equal in pure water, a mathematical expression can be written:

$K_w = [H^+][OH^-]$ for pure water at 25°C

$K_w = [1.00 \times 10^{-7}][1.00 \times 10^{-7}] = 1.00 \times 10^{-14}$

At 25°C, for any aqueous solution, $K_w = 1.00 \times 10^{-14}$. This is a very useful relationship, and will be used in later calculations.

V. pH

Now that we know that $[H^+]$ and $[OH^-]$ ion concentrations are always equal to 1.00×10^{-7} M for pure water, the next step is to convert this into the pH of pure water. It is a simple task because all we need to do is to use the formula $pH = -\log[H^+]$ and the $[H^+]$ for pure water.

$pH = -\log[1.00 \times 10^{-7}]$

$pH = -(-7.0000)$

$pH = 7.000$

When small amounts of **acid** are added to pure water, the acid adds to the total amount of $[H^+]$ available in solution, and therefore changes the pH of the solution. For all practical purposes, even in very dilute acid solutions, the concentration of the acid can be used to calculate the total $[H^+]$.

Example: To calculate the pH of a 0.002 M solution of the strong acid, HCl, you would need to write an equation for its ionization.

0.002 M   0.002 M   0.002 M

$$HCl \longrightarrow H^+ \quad + \quad Cl^-$$

if $[H^+] = 0.002$, then

$[H^+] = 2.00 \times 10^{-3}$, and then pH would be

$pH = -\log [2.00 \times 10^{-3}]$

$pH = -(-2.69897) = 2.70$ (rounded)

1. For each of the following $[H^+]$, calculate the solution's pH. Round off all pH numbers to two numbers after the decimal.

   a. $[H+] = 2.35 \times 10^{-4}$ pH = _____

   b. $[H+] = 9.67 \times 10^{-1}0$ pH = _____

   c. $[H+] = 1.00 \times 10^{-6}$ pH = _____

   d. $[H+] = 3.89 \times 10^{-2}$ pH = _____

   e. $[H+] = 0.00000258$ pH = _____

VI. pH and pOH

In a fashion similar to pH, chemists have also devised an expression for $[OH^-]$, which is the pOH. The mathematical formula is also similar:

$$pOH = -\log [OH^-]$$

The pH and pOH scales are related to the ionization constant of water, K_w. You will remember that at 25°C, $K_w = [H^+][OH^-]$. In pure water both the $[H^+]$ and the $[OH^-]$ are equal to 1.00×10^{-7}. Therefore, if these values are placed into the same equation, the following results can be obtained:

$$[H^+][OH^-] = 1.00 \times 10{-14}$$

Multiplying the above equation by $-\log$ gives:

$$pH + OH = 14.$$

This useful relationship makes determining the pH of basic solutions a much simpler task.

Example: To calculate the pOH of a 0.002 M solution of a strong base, KOH, you would need to write an equation for its ionization.

0.002 M   0.002 M   0.002 M

$$KOHl \longrightarrow K^+ \quad + \quad OH^-$$

if $[OH^-] = 0.002$, then

$[OH^-] = 2.00 \times 10^{-3}$, and then pOH would be

$pOH = -\log[2.00 \times 10^{-3}]$

$pOH = -(-2.69897) = 2.70$ (rounded)

It may be noted that these are exactly the same steps taken in the previous example for pH, except this time it is $[OH]$ and the results are for pOH. To complete the calculation of pH, only one step is necessary:

if $pH + OH = 14$, then when $pOH = 2.70$

$pH = 14 - 2.70 = 11.70$

The relationship between pH and pOH can be summarized as follows:

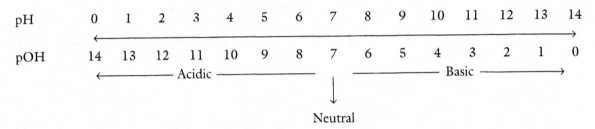

From this scale it can be seen that when the pH is from 0 – 6.99 the solution is acidic, and if the pH is from 7.01 – 14 the solution is basic. In contrast, the pOH scale numerical values run in the opposite direction.

1. For each of the following $[OH^-]$, calculate the pOH of the solution. Round off all numbers to two places after the decimal.

    a. $[OH^-] = 2.35 \times 10{-3}$ pOH = _____

    b. $[OH^-] = 6.89 \times 10{-13}$ pOH = _____

    c. $[OH^-] = 1.00 \times 10{-6}$ pOH = _____

    d. $[OH^-] = 7.55 \times 10{-4}$ pOH = _____

    e. $[OH^-] = 8.62 \times 10{-10}$ pOH = _____

2. Find the pH of each of the following solutions if their pOH is:

    a. pOH = 3.00   pH = _____

    b. pOH = 5.50   pH = _____

    c. pOH = 8.25   pH = _____

    d. pOH = 13.50   pH = _____

    e. pOH = 1.55   pH = _____

VII. pH to [H⁺] and pOH to [OH⁻]

Now there is only one more thing to learn to complete our study of the pH of solutions: how to convert pH or pOH back into their respective hydrogen or hydroxide ion concentration. Again, the process is easy with the aid of the electronic calculator.

Example: If by means of a pH meter or some other method it was determined that a solution had a pH = 6.50. What would be the [H⁺] of the solution?

The mathematical process we are going to need to do is called finding the **antilog**. This means doing the opposite of finding the log. The steps to do this are the following:

Enter the pH number "6.50"; then, push the "±" key followed by the "INV" key; and finally, the "LOG" key. The resulting answer, 3.16×10^{-7} (rounded), would be the [H⁺] of the solution.

The key sequence is the same, whether you are converting from pH to [H⁺] or from pOH to [OH⁻].

VIII. Summary Diagram

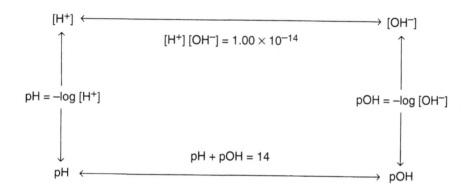

$$[H^+][OH^-] = 1.00 \times 10^{-14}$$

$$pH = -\log[H^+] \qquad pOH = -\log[OH^-]$$

$$pH + pOH = 14$$

Using the above diagram as a guide, complete the following:

1. Determine the [H⁺] of the following solutions using the pH values given:

    a. pH = 7.00   [H⁺] = _____

    b. pH = 2.45   [H⁺] = _____

    c. pH = 12.50   [H⁺] = _____

    d. pH = 4.65   [H⁺] = _____

    e. pH = 9.85   [H⁺] = _____

2. What is the pH and pOH of a solution with [H⁺] = 0.00015?

                                                          pH = _____

                                                          pOH = _____

3. What is the pH and pOH of a 0.01 M NaOH solution?

pH = _____

pOH = _____

4. What is the pH and pOH of a 0.01 M HCl solution?

pH = _____

pOH = _____

5. If the $[H^+]$ of a solution is 3.50×10^{-3},

a. What is the solution's pH?   pH = _____

b. What is the solution's pOH?   pOH = _____

c. What is the $[OH^-]$? $[OH^-]$ = _____

d. Is the solution acidic or basic?   _____

22 – pH and pOH

Possible Points = 20

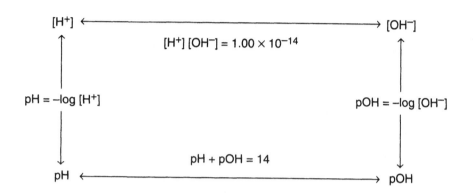

I. On each line of the table below, one piece of information has been provided. Using the diagram above as a guide, do the necessary calculations to make all blanks on each line agree.

$[H^+]$	$[OH^-]$	pH	pOH	Acid, Base or Neutral
1.0×10^{-4}				
		8.50		
			7.00	
	4.5×10^{-9}			
7.5×10^{-12}				

Name: _____

Date: _____ Section: _____

23 – Energy

Possible Points = 20

$$KE = \tfrac{1}{2}mv^2 \qquad U_{grav} = m\,p\,h \qquad power = \frac{energy\ transferred}{time}$$

$$1\ watt = 1\ J/s \qquad watts = amperes \times volts \qquad 1\ calorie = 4.2\ J$$

1. What is the kinetic energy, in joules, of a 10.00 g bullet traveling at 1,000 m/s?

2. The kinetic energy of a moving object is 250 J. If the object has a mass of 5.0 kg, how fast is it moving?

3. What is the gravitational potential energy of a 700 kg grand piano at the top of a 400 m building?

4. A 60 kg object falls from a height of 2.0 m.

   a. What was the gravitational potential of the object before falling?

   b. What is its kinetic energy the instant it hits the ground?

   c. What is its velocity when it strikes the ground?

5. An electric heater has a resistance of 8 ohms and draws 20 amperes. Determine the rate at which heat is produced. What is the cost of operating the heater for 2 hours if the cost of the electricity is 15.4¢/kw · hr.

24 – Waves and Oscillations

Possible Points = 20

$$T = 2\pi\sqrt{\frac{m}{k}} \qquad f = \frac{1}{T} = \frac{1}{2\pi}\sqrt{\frac{k}{m}} \qquad v_{\text{wave}} = \frac{\lambda}{T} \qquad v_{\text{wave}} = \lambda f$$

1. If a wave has a frequency of 5.00×10^8 Hz and a velocity of 200 m/s, what is its wavelength?

2. What is the period, T, of a wave that has a frequency of 5.00×10^8 Hz?

3. The speed of sound in air at 0°C is 331 m/s. Compute the speed in air at 24°C, using the approximate correction of 60 cm/s/°C.

4. The FM band on the radio is centered at a frequency of 100 megacycles. Assuming the waves travel at 3×10^8 m/s, what is the length of an FM antenna arm if it is one-quarter wavelength?

5. What is the period, T, of a 60 kg bungee jumper that is supported by a rubber band that has a "spring constant," $k = 35$ N/m?

25 – Ohm's Law 1

Possible Points = 20

$$V = I R$$

1. If a 1.5 volt dry cell is placed in a circuit with a 0.50 Ω resistor, how much current will flow through the resistor?

2. If an ammeter connected in series with an unknown resistor reads 0.50 ampere and a voltmeter placed across the ends of the resistor reads 1.5 volts, what is the value of the resistor?

3. A car battery in a closed circuit has a terminal voltage of 12.00 volts and is connected in series with a rheostat and an ammeter that reads 5.00 amps. Neglecting the battery's internal resistance and the resistance of the ammeter, what is the resistance of the rheostat?

4. In a circuit, a current of 15.0 amperes divides between two parallel resistors of 8 Ω and 3 Ω, respectively.

   a. What is the equivalent resistance of the parallel combination?

   b. What will be the voltage across the parallel combination?

   c. What will be the current in the 8 Ω resistor?

   d. What will be the current in the 3 Ω resistor?

5. A motor is designed to operate at a current of 3.50 amps when the voltage is 115 volts. How large of a resistor would have to be placed in series with the motor to maintain this current at 130 volts?

26 – Ohm's Law 2

Possible Points = 20

$$V = I R$$

1. A 5 Ω resistance is in series with a light bulb and connected to a 120 volt source. Determine the resistance of the light bulb, if the current in the circuit is 3.0 amperes.

2. A closed circuit is formed by a uniform 200 cm wire that has a total resistance of 10.0 Ω and that is connected in series with a 12.00 volt battery and a rheostat that has a resistance of 1.00 Ω. What is the reading of a voltmeter that is placed across 50 cm of the wire?

3. A 40 Ω galvanometer is shunted by a 4 Ω wire. What percent of the total current will flow through the galvanometer and what percent through the shunt?

4. It is desired to have 20% of the total current pass through an ammeter that has a resistance 0.05 Ω. What would be the resistance of the shunt that must be placed in parallel with it?

5. Three resistors of 2 Ω, 4 Ω, and 6 Ω are connected in parallel with 12 volts across the combination. Determine the current in each resistor and the total current in the system.

Name: _____

Date: _____ Section: _____

27 – Refraction

Possible Points = 20

$$n_1 \sin \theta_1 = n_2 \sin \theta_2 \qquad n_{material} = \frac{\text{speed of light in vacuum}}{\text{speed of light in the material}}$$

1. A light ray strikes a glass surface at an incident angle of 55°. The refractive index of the glass is 1.5.

   a. What is the direction of the reflected ray?

   b. What is the direction of the refracted ray?

2. Assume that the index of refraction for air and a vacuum are the same. What is the refractive index for water, if the speed of light in water is 0.75 that in air?

3. What is the angle of refraction for a light ray that is traveling in a medium with an index of refraction of 1.2, when it enters a medium with an index of refraction of 1.5 at an angle of 17°?

4. What is the critical angle, θ_c, between air ($n_t = 1.00$) and water ($n_i = 1.33$), when light traveling through the water encounters the water/air interface? (Hint: $\sin \theta_c = n_t / n_i$)

5. A light ray is incident upon a flat piece of glass ($n = 1.5$). The angle of incidence is 50° from the normal.

   a. What is the angle of the reflected ray?

   b. What is the angle of the refracted ray?

28 – Diffraction

Possible Points = 24

$$n\lambda = d \sin \theta$$

1. If the angle of the first-order diffraction of the 530 nm line is 32.0° when viewed through a diffraction grating, what must be the spacing (cm) between the grating lines?

2. Red light strikes a diffraction grating ruled with 4,000 lines/cm and it produces a second order image that is diffracted 32.0° from the normal. What is the wavelength of the light?

3. The most intense yellow line in the first order ($n = 1$) spectrum of sodium metal is called the D line. What is its wavelength (nm) if it appears at an angle of 36.0° from normal, when viewed through a diffraction grating that has 10,000 lines/cm?

4. Green light, with a wavelength of 540 nm, strikes a diffraction grating with 2,500 lines/cm. What is the angle of the third-order image formed?

5. If a diffraction grating with lines 0.00060 cm apart is illuminated by a light with a wavelength of 550 nm, at what angle will the second-order line appear?

6. The second- and third-order visible spectra of a diffraction grating partially overlap. What $n = 3$ wavelength will appear at the $\lambda = 700$ nm position of the $n = 2$ spectrum?

Name: _____

Date: _____ Section: _____

29 – Thin Lenses

Possible Points = 20

$$\frac{1}{f} = \frac{1}{s} + \frac{1}{s'} \qquad \text{magnification} = \frac{\text{image size}}{\text{object size}} = -\frac{\text{image distance}}{\text{object distance}} = -\frac{s'}{s}$$

1. What is the focal length of a converging lens, if the object and image distances are 90 cm and 50 cm respectively?

2. If the focal length of the lens in an old portrait camera is 25 cm, how far behind the lens should the film be placed to record the image of a person seated 1.3 m in front of the lens?

3. An object is 10 cm high and is placed 30 cm in front of a concave lens with a focal length of –18 cm. What is the position, the nature (virtual or real), and the height of the image formed?

4. A small object is placed 18 cm in front of a converging lens with a focal length of 12 cm. Find the position, the nature (virtual or real), and the magnification of the image formed.

5. As shown in the figure, an optical system is composed of two lenses that are separated by 24 cm and have focal lengths of 4.5 cm and 7.0 cm respectively. They form an image on a screen placed 20 cm behind the second lens.

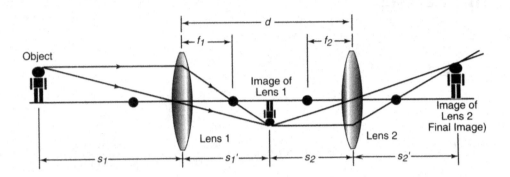

  a.   Where does the image from the first lens form?

  b.   How far in front of the first lens would the object have to be placed?

30 – Organic Structures and IUPAC Names

Possible Points = 15

I. Isomers – by definition, have identical molecular formulas but different structural formulas. Indicate if the following pairs of compounds are the same [S], isomers [I], or unrelated [U].

1. 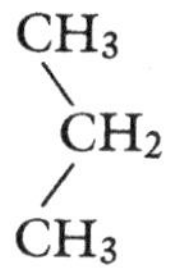
$$CH_3$$
$$CH_2$$
$$CH_3$$

$$CH_3-CH_2-CH_3$$

—————— [   ]

2. $$CH_3-CH-CH_3$$
$$CH_3$$

$$CH_3-CH \begin{smallmatrix} CH_3 \\ CH_3 \end{smallmatrix}$$

—————— [   ]

3. $$\overset{O}{\overset{||}{CH_3-CH_2-C-O-H}}$$

$$\overset{O}{\overset{||}{H-O-C-CH_3}}$$

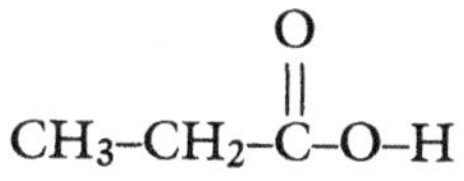

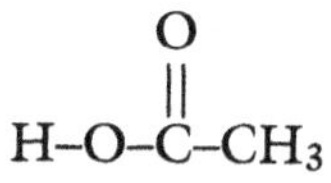

—————— [   ]

4. $$CH_3-CH=CH-CH_3$$

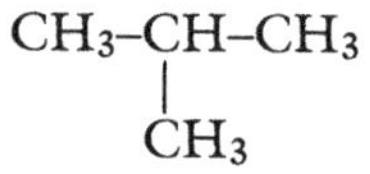

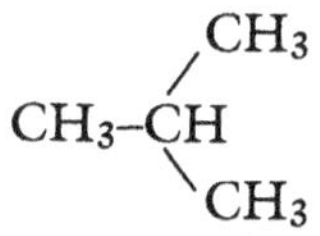

$$\overset{O}{\overset{||}{CH_3-CH_2-C-N-H}}$$
$$H$$

—————— [   ]

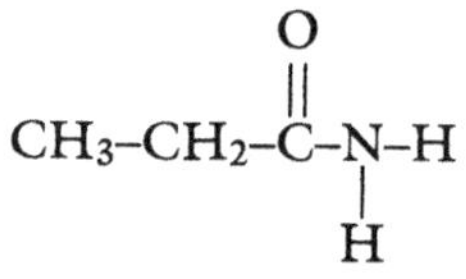

5. $$CH_3-CH=CH-CH_3$$

$$CH_2=CH-CH_2-CH_3$$

—————— [   ]

II. When using the IUPAC naming system, any carbon chain that cannot be included as part of the main (longest) chain must be named as a side chain. The following side chains (alkyl groups) are commonly used as a part of the IUPAC naming system.

←———————————————— longest carbon chain ————————————————→

|
CH_3 CH_2 CH_2 CH–CH_3 CH_2 CH_3–CH_2–CH–CH_3 CH_3–C–CH_3 CH_2–CH–CH_3
          CH_3 CH_2 CH_3 CH_2 CH_3 CH_3
                   CH_3 CH_2
                                   CH_3

methyl   ethyl   propyl   isopropyl   butyl   sec. butyl   t-butyl   isobutyl

(If any sequence of carbon atoms that is 3 or longer are connected into a ring, the prefix cyclo- is added. Examples are cyclopropyl-, cyclobutyl-, etc.)

III. Give each of the following compounds a correct IUPAC name.

1. CH_3–CH_2–CH–CH_2–CH–CH_3 _____
                   CH_3 CH_3

2. CH_3 CH_2–CH_3
        CH_3–CH_2–CH–CH –CH _____
        CH_3–CH_2–CH–CH_3 CH_3

3. CH_2–CH_3
        CH_3–CH_2–C–CH_2–CH_2–CH_2–CH_3 _____
              CH_2–CH_2–CH_3

4. CH_2–CH_2
        CH_2–CH–CH_2–CH_2–CH_3 _____

5. CH_2–CH_2–CH_3
        CH–CH_2–CH_2–CH_3 _____
        CH_2–CH_2–CH_2–CH_3

IV. In the space provided, draw the correct structure formula for each of the following compounds.

1. 3-methylhexane

2. 2,3-dimethyl-3-ethylhexane

3. 2-ethyl-1-isopropylcyclohexane

4. 2-methyl-5-sec. butyloctane

5. 3,4-diethyl-4-propyldecane

Name: _____

Date: _____ Section: _____

31 – Naming Alkanes

Possible Points = 15

I. Draw structural formulas for the following compounds.

1. 1-chloro-3-ethyl-2-methylhexane

2. 2,4,5-tribromo-4-isopropyl-2-methylnonane

3. 3-ethyl-1,1-dimethylcyclopentane

4. 1,1,1-trifluro-3-sec. butyldecane

5. 2-cyclopropylpropane

II. Give each of the following compounds a correct IUPAC name.

1. 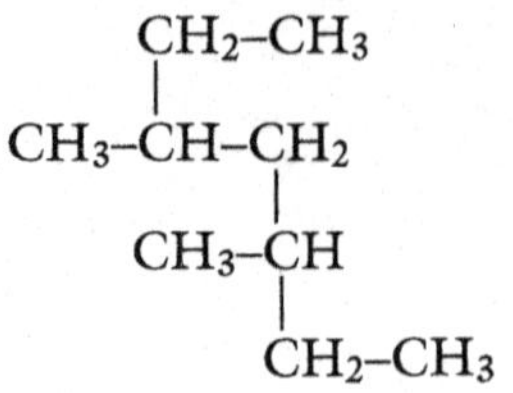
$$CH_2-CH_3$$
$$CH_3-CH-CH_2$$
$$CH_3-CH$$
$$CH_2-CH_3$$

_____

2. $CHCl_3$

_____

3. CCl_4

_____

4. 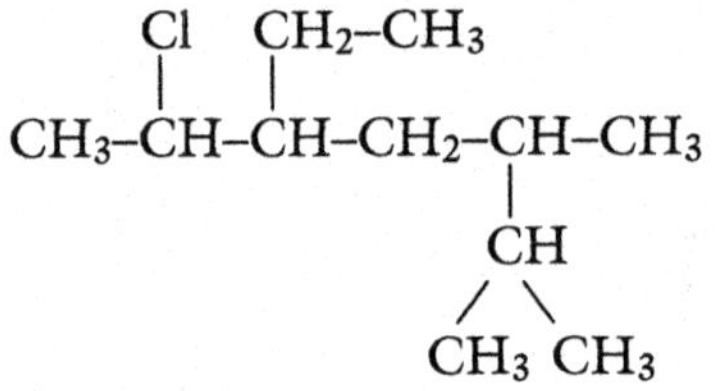
Cl CH_2-CH_3
$CH_3-CH-CH-CH_2-CH-CH_3$
CH
CH_3 CH_3

_____

5. CCl_2F_2

_____

6.

_____

7. 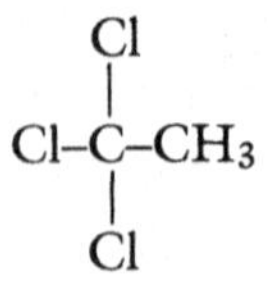
Cl
$Cl-C-CH_3$
Cl

_____

8. 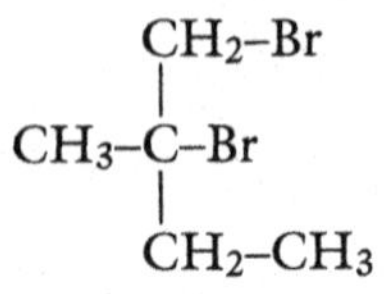
CH_2-Br
CH_3-C-Br
CH_2-CH_3

_____

9. CH_3I

_____

10. CH_3Br

_____

Exercise 32 - Alkane Names and Reactions

Name: ______________________

Date: ____________ **Section:** ________

32 - Alkane Names and Reactions

Possible Points = 15

I. **Draw structural formulas for the following compounds.**

1. 3-isopropylhexane

2. 2,4-dichloropentane

3. 3-ethyl-2,4,3-trimethyloctane

4. t-butylcyclohexane

5. 1-bromo-2-methylcyclobutane

II. Give each of the following compounds a correct IUPAC name.

1. 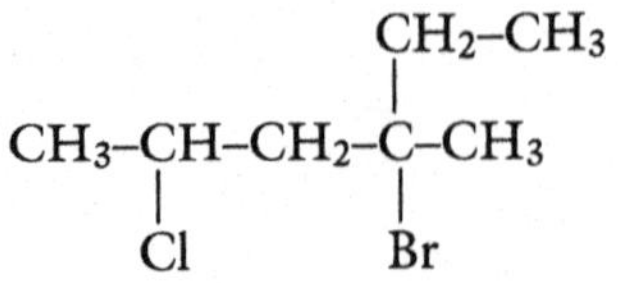
$$CH_3-CH-CH_2-C-CH_3$$
with CH_2-CH_3 above the C, Cl below the first CH, Br below the C

_____

2.
$$CH_3-CH-CH-CH_2-CH_2$$
with Br and F below, CH_3 below the last CH_2

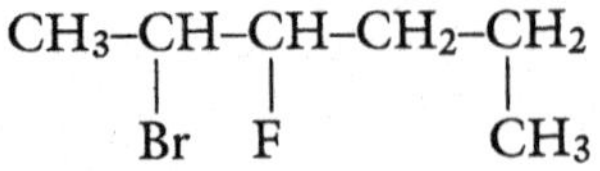

_____

3. CH_3-CH_3

_____

4. 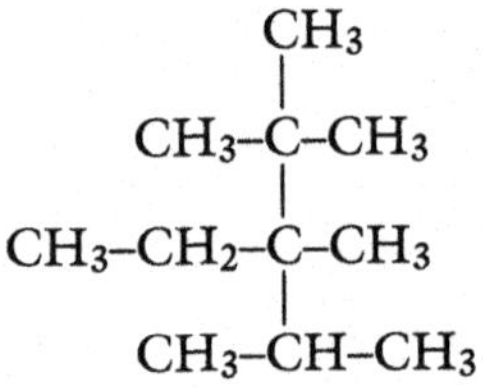
$$CH_3-C-CH_3$$ with CH_3 above
$$CH_3-CH_2-C-CH_3$$
$$CH_3-CH-CH_3$$

_____

5. 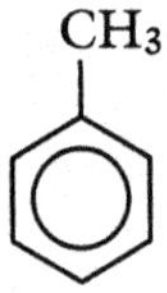   benzene ring with CH_3

_____

III. Complete and balance the following equations.

1. $CH_4 + O_2 \longrightarrow$

_____

2. $CH_3-CH_2-CH_3 + O_2 \longrightarrow$

_____

3. $CH_3-CH_3 + H_2SO_4 \longrightarrow$

_____

4. $CH_4 + Br_2 \xrightarrow[200-400°C]{u.v.}$

_____

5. $CH_4 + NaOH \longrightarrow$

_____

Common Temperature Measurements

Temperature	Kelvin	Celsius	Rankine	Fahrenheit
Absolute zero	0 K	−273.15°C	0 R	−459.67°F
Freezing point of oxygen, O_2	54.8 K	−218.4°C	98.5 R	−361.1°F
Freezing point of nitrogen, N_2	63.3 K	−209.86°C	113.9 R	−345.75°F
Melting point of water	273.15 K	0°C	491.67 R	32°F
Boiling point of water	373.15 K	100°C	671.67 R	212°F
Melting point of lead	600.65 K	327.5°C	1,081.17 R	621.5°F
Melting point of Iron	1,808 K	1,535°C	3,255 R	2,795°F
Conversion, Celsius to Fahrenheit	°F = 9/5°C + 32			
Conversion, Fahrenheit to Celsius	°C = 5/9(°F − 32)			
Conversion, Celsius to Kelvin	K = °C + 273.15			
Conversion, Fahrenheit to Rankine	R = °F + 459.67			

Prefixes Used with SI Fundamental Units

Prefix	Multiple or Fraction	Scientific Notation	Pronunciation
G	1,000,000,000	10^9	giga
M	1,000,000	10^6	mega
k	1,000	10^3	kilo
h	100	10^2	hecto
da	10	10^1	deka
*	1.	10^0	*
d	0.1	10^{-1}	deci
c	0.01	10^{-2}	centi
m	0.001	10^{-3}	milli
μ	0.000001	10^{-6}	micro
n	0.000000001	10^{-9}	nano
p	0.000000000001	10^{-12}	pico

Appendix

C

Derived Units

Examples of SI Derived Units		
Derived Quantity	**SI Derived Unit**	
	Name	**Symbol**
Area	square meter	m^2
Volume	cubic meter	m^3
Speed, velocity	meter per second	m/s
Acceleration	meter per second squared	m/s^2
Wave number	reciprocal meter	m^{-1}
Mass density	kilogram per cubic meter	kg/m^3
Specific volume	cubic meter per kilogram	m^3/kg
Current density	ampere per square meter	A/m^2
Magnetic field strength	ampere per meter	A/m
Amount of substance concentration	mole per cubic meter	mol/m^3
Luminance	candela per square meter	cd/m^2
Mass fraction	kilogram per kilogram, which may be represented by the number 1	kg/kg = 1

SI Derived Units with Special Names and Symbols				
Derived Quantity	**SI Derived Units**			
	Name	**Symbol**	**Expression in other SI Units**	**Expression in SI Base Units**
Plane angle	radian [a]	rad	–	$m \cdot m^{-1} = 1$ [b]
Solid angle	steradian [a]	sr [c]	–	$m^2 \cdot m^{-2} = 1$ [b]
Frequency	hertz	Hz	–	s^{-1}
Force	newton	N	–	$m \cdot kg \cdot s^{-2}$
Pressure, stress	pascal	Pa	N/m^2	$m^{-1} \cdot kg \cdot s^{-2}$
Energy, work, quantity of heat	joule	J	$N \cdot m$	$m^2 \cdot kg \cdot s^{-2}$
Power, radiant flux	watt	W	J/s	$m^2 \cdot kg \cdot s^{-3}$
Electric charge, quantity of electricity	coulomb	C	–	$s \cdot A$
Electric potential difference, electromotive force	volt	V	W/A	$m^2 \cdot kg \cdot s^{-3} \cdot A^{-1}$
Capacitance	farad	F	C/V	$m^{-2} \cdot kg^{-1} \cdot s^4 \cdot A^2$
Electric resistance	ohm	W	V/A	$m^2 \cdot kg \cdot s^{-3} \cdot A^{-2}$
Electric conductance	siemens	S	A/V	$m^{-2} \cdot kg^{-1} \cdot s^3 \cdot A^2$
Magnetic flux	weber	Wb	$V \cdot s$	$m^2 \cdot kg \cdot s^{-2} \cdot A^{-1}$
Magnetic flux density	tesla	T	Wb/m^2	$kg \cdot s^{-2} \cdot A^{-1}$
Inductance	henry	H	Wb/A	$m^2 \cdot kg \cdot s^{-2} \cdot A^{-2}$
Celsius temperature	degree Celsius	°C	–	K
Luminous flux	lumen	lm	$cd \cdot sr$ [c]	$m^2 \cdot m^{-2} \cdot cd = cd$
Illuminance	lux	lx	lm/m^2	$m^2 \cdot m^{-4} \cdot cd = m^{-2} \cdot cd$
Activity (of a radionuclide)	becquerel	Bq	–	s^{-1}
Absorbed dose, specific energy (imparted), kerma	gray	Gy	J/kg	$m^2 \cdot s^{-2}$
Dose equivalent [d]	sievert	Sv	J/kg	$m^2 \cdot s^{-2}$

(a) The radian and steradian may be used advantageously in expressions for derived units to distinguish between quantities of a different nature but of the same dimension.

(b) In practice, the symbols rad and sr are used where appropriate, but the derived unit "1" is generally omitted.

(c) In photometry, the unit name steradian and the unit symbol sr are usually retained in expressions for derived units.

(d) Other quantities expressed in sieverts are ambient dose equivalent, directional dose equivalent, personal dose equivalent, and organ equivalent dose.

This information taken from the Foundation of Modern Science and Technology from the Physics Laboratory of NIST: http://physics.nist.gov/cuu/Units/index.html.